# M60 *PATTON* in Action

by Jim Mesko

color by Don Greer

illustrated by Perry Manley

M609A1 of the 3rd Battalion, 32nd Armor, 3rd Armored Division, 7th Army, during the Fall 1977. Scheme is Sand, Red Brown, Green, and Black, with all markings in Black.



1115 CROWLEY DRIVE, CARROLLTON, TEXAS 75011-5010



**ISBN 0-89747-176-8**

**If you have any photographs of the aircraft, armor, soldiers or ships of any nation, particularly wartime snapshots, why not share them with us and help make Squadron/Signal's books all the more interesting and complete in the future. Any photograph sent to us will be copied and the original returned. The donor will be fully credited for any photos used. Please indicate if you wish us not to return the photos. Please send them to: Squadron/Signal Publications, Inc., 1115 Crowley Dr., Carrollton, TX 75011-5010.**

## Photo Credits

US Army
US Marine Corps
Mike Green
Saito Wakui
Chrysler Corporation
Seventh Army, Europe
Fort Knox PIO
Teledyne Continental Motors
MERADCOM
3rd Armor Division
1st Armor Division
Robert Icks
Jim Mesko
Major Gordan Kurtz
Gary Binder
Ohio National Guard, UTES 1
Sergeant 1st Class Gary D Accord
Staff Sergeant G P Mauerman
Sergeant D LeMasters
CWO3 Ray Conti
Staff Sergeant Daniel Peterson
1st Infantry Division
Rock Island Arsenal
4th Infantry Division
Israeli Defense Force
Israeli Embassey
Egyptian Embassey
US Air Force
W E Storey

**(Right) American armored power. A column of M60s passes in review during ceremonies in Germany in 1966 honoring the 47th anniversary of the 3rd Infantry Division. The tanks are from the 3rd Battalion of the 64th Armor. These tanks can ultimately trace their lineage back to the M26 Pershing which initially saw service in 1945, a longevity which few other weapons systems can claim. (US Army via Binder)**

3·3△64

# INTRODUCTION

When World War II ended, the United States possessed one of the largest armored forces in the world. The main battle tank of this force, the M4 Sherman was a robust and mechanically reliable vehicle, which had been introduced into service during the middle of 1942. Unfortunately by 1945, despite some upgrading of both its armor and armament, the M4 was at a decided disadvantage when it met German Tigers and Panthers.

Belatedly, Army officials tried to make up for lost time, pushing development of a number of tank designs which could hold their own against the heavier German tanks. By early 1945, one of these new designs, the M26 Pershing Heavy Tank was judged ready for field use, and a number were rushed to Europe to help shore up hard pressed American tankers. The M26 Pershing did much to redress the balance between American and German armor, and performed extremely well in the few encounters with German tanks which occurred before the war ended.

With Germany's surrender Army officials were able to examine the latest German tanks and designs under development. At the same time most of the latest Soviet tanks, many of which totally outgunned anything in the US or allied inventory, were examined. These examinations had a very sobering affect on American armored officers. As a result, attempts were made to upgrade the M26 Pershing, and to develop a successor to it, so as not to fall behind the Russians.

However, for a number of reasons tank development was given a low priority in the years immediately following the end of World War II. Chief among these was the availability of the atomic bomb, which some military and political leaders felt would render conventional weapons such as the tank obsolete. Since the main delivery system for this new weapon was the airplane the Air Force became the nation's chief military force, receiving a disproportionate share of the defense budget. Another factor which hindered the development of new American armored vehicles was the belief on the part of many US officials, and much of the American public, that our wartime Russian allies would be interested in keeping the peace. In the euphoria that followed the defeat of the Axis Powers, few people believed that our former ally had any desire other than to live peaceably after the unbelievably bitter experiences they suffered in the war. Unfortunately the communist mentality was underestimated, and the Soviets quickly began setting up brutal puppet states in eastern Europe and the far east. Hostility immediately began to surface in various parts of the world.

Belatedly, efforts were made to upgrade the Pershing which suffered from both engine and trans—mission problems. The end result was the M46 which was basically an M26 Pershing with a new AV-1790-5A engine and CD-850 transmission which formed a drive train which would power the US main battle tank for the next thirty years. Changes were made to the fire control system, and a bore evacuator was added to the 90MM gun barrel. These rebuilt tanks were nicknamed 'Patton' in honor of the most famous World War II American armor commander.

While this program did result in a general upgrading of the American armored capability, it was a stop-gap measure. Realizing this, the army authorized a design program for an interrelated family of three tanks in the light, medium, and heavy classes. Designated respectively the T41, T42, and T43 with all three tanks sharing many common components in order to reduce logistical requirements in time of war. However, work on these tank designs progressed slowly due to fiscal restrictions, peacetime apathy, and the still unanswered question of the value that tanks would have in future conflicts.

This complacency, however, was suddenly shattered on the morning of 25 June 1950 when 150 Russian made T34s of the North Korean Army spearheaded a massive attack into South Korea*. The North Koreans quickly routed the South Koreans who had no tanks and few weapons capable of stopping the T34. American troops, rushed to the battlefield from occupation duty in Japan without tank support faired little better. Eventually a few M4 Shermans and M26 Pershings were finally committed to the fighting and helped to blunt the communist drive. As the war progressed more Shermans and Pershings were committed, as well as the newer M46 Patton, and the North Koreans were eventually driven back across the original border deep into North Korea.

When the North Koreans launched their surprise attack, the T42 project was still in the working up process, and while work on the project was pushed ahead, the new vehicle proved to be underpowered and was deemed unsuitable for use in the field. However, the new T-42 turret had improved ballistic protection, a better internal layout, and superior rangefinding equipment compared to the M46 Patton. The army decided to mount this turret on a modified M46 chassis under the designation M47. Much was expected of this new combination but unfortunately the haste in which it was pushed through, pushed through without proper testing, resulted in numerous problems. The M47 saw no action in the Korean War where the Sherman, Pershing, and Patton shouldered the fighting until a cease-fire was declared during the summer of 1953.

The M47 finally went into service during 1952 when units of the 1st and 2nd Armored Divisions began receiving the new tank. While technically superior to the Patton, early teething problems of the M47 caused much concern in the army, both at the field level and in higher circles. At the same time a totally new tank was under development, and by 1955 this vehicle, the M48 was ready for issue to field units. As a result the M47 was declared *limited standard* and quickly phased out of US service, with the bulk of the 8000 M47s produced, being supplied to American allies around the world. Over 7000 M47s were supplied to NATO countries where they formed the backbone of Western European

**For a history of armor in the Korean War consult the author's ARMOR IN KOREA also published by Squadron Signal.*

**(Below) The M4 Sherman was the backbone of US armed forces during the Second World War and into the Koren War. This M4A3 equipped with a dozer blade sits outside of Seoul during the see-saw fighting which took place after Chinese forces entered the Korean War. (US Army)**

**(Below) The replacement for the M4 Sherman was the M26 Pershing which saw service during the waning days of World War II in Europe. It proved to be far more effective against the heavier German tanks than did the Sherman. This Marine Corps Pershing covers the Naktong River during the later stages of fighting around the Pusan Perimeter (USMC)**

armored forces for over fifteen years. Because of its widespread usage the M47 saw action in a number of conflicts including the Suez Crisis (1956), the Indo-Pakistani War (1965), The Arab-Israeli War (1967), the Cyprus Invasion (1974), the Spanish Sahara (1974), and the Ogaden Desert War (1977). Although still in service with a number of smaller nations as of this date, the M47 has reached the end of its active service life and the next few years will see it being replaced by more modern tanks.

While the M47 program was viewed as a quick way to redress the imbalance between US and Russian armor, Army officials realized that it was an interim measure and a completely new design would be needed to forestall future imbalance. Accordingly, the Detroit Arsenal began studying ways in which to improve the basic design of the M47 in October of 1950. By December of 1950 the study was completed and after evaluation the army decided to order the new design. A contract was awarded to Chrysler Corporation the same month for six prototypes of the new design under the test designation T48. By the following December the test vehicles were ready for work-up, which began in 1952.

The new tank, under the production designation M48 and also christened the 'Patton', began entering service in 1953. Unfortunately, as with the M47 the army's desire to put the new vehicle into production and service also resulted in an inadequate testing and work-up period. Numerous technical problems arose on production machines which could have been eliminated if an adequate amount of time been allocated to testing and work up. Thus most of the early production M48s required extensive work to make them operational. These early machines also had an exceedingly high rate of fuel consumption which necessitated the addition of four fifty-five gallon fuel drums being carried on a jettisonable rack behind the engine compartment. The M48 was quickly followed by the M48A1 which differed only in that a machine gun cupola was fitted to the turret top in place of the exposed machine gun mount of the tank commanders hatch.

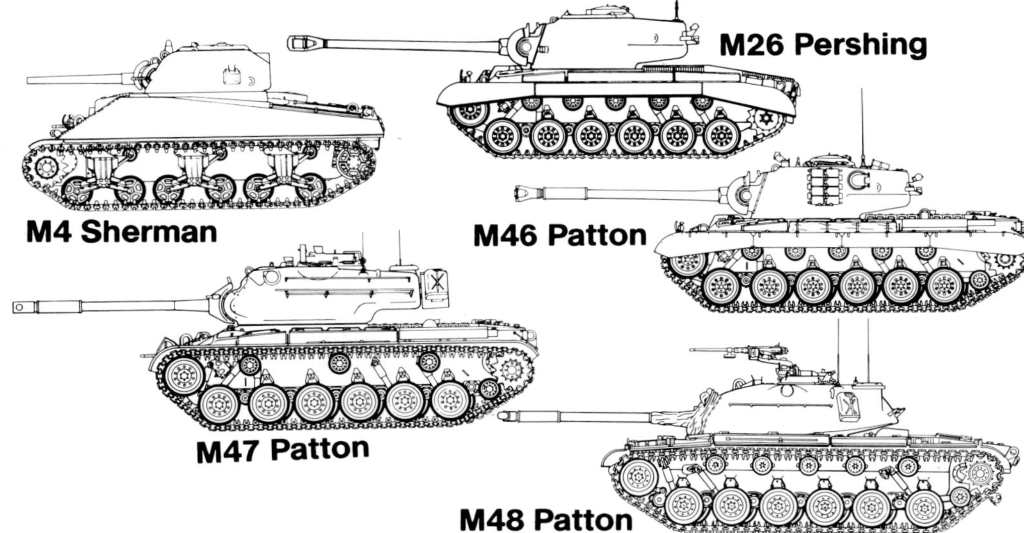

**(Above) While satisfied in general with the M26 the army decided to modernize the Pershing and correct certain mechanical problems which the tank exhibited. The rebuilt Pershing was designated the M46 Patton and featured a new transmission, engine, bore evacuator, as well as changes in the fire control and suspension systems.**

**(Below) The next major American tank was the M47 which resulted from the grafting of the turret from the cancelled T42 series onto the hull of an M46. This South Korean Marine M47 taking part in *Exercise Team Spirit 1983* carries the unusual camouflage pattern often found on Korean tanks. (Wakui via Green)**

Though relatively pleased with the new tank once the bugs were worked out, the Pattons short range resulted in a new variant being put into production under the designation M48A2 which featured an improved fuel injected engine, a redesigned rear hull and more internal fuel capacity. This new model, introduced in 1955, had over twice the range of its predecessors and was the first model to incorporate the various modifications made to the early models. As such it was the first production Patton issued to the army which did not need extensive modifications to make it operational.

By the late 1950s, production of the M48 had ceased as the army shifted over to production of the new M60, a much improved and modified version of the basic Patton design. However, since there were still many M48s and M48A1s still in service, the decision was made to use M60 components to rebuild these older vehicles. These rebuilt vehicles, re-designated the M48A3, featured a new AVPS-01790 diesel engine, and the improved fire control system from the later model M48A2. Many of these rebuilt Pattons saw combat in Vietnam where they proved to be a rugged, robust and reliable vehicle.

**(Below) This M48 is being demonstrated to the press during ceremonies at Aberdeen Proving Grounds. (US Army)**

As the armament of the Russian main battle tanks increased, the army looked at the possibility of up gunning the M48. The Israelis had rearmed their M48s in the late sixties with a 105MM cannon with excellent results. Although usually referred to with the US service designation of M48A4 for administrative and identification purposes by the US Army, these up-gunned Israeli tanks were never a US service model.

Following the Yom Kipper War in 1973, US Army tank inventory dropped to dangerously low levels as large numbers of M60s were shipped to Israel as replacements for the losses she suffered. Problems with the production version of the M60 series forced the army to look for a means by which US tank strength could rapidly be built up. The Pentagon decided to modernize over 1,600 M48s with M60 components, an armament upgrading by replacing the 90MM gun with a 105MM gun, re-working the fire control system, and replacing the engines on the earlier models. Under the designation M48A5, this variant compared favorably in most respects with the M60 and provided a relatively quick and inexpensive way to build up US tank inventory. While the only regular army tank units to receive the M48A5 were stationed in Korea, the M48A5 became the backbone of National Guard armored units. Several hundred were also supplied to foreign countries, and a conversion package was made available to those US allies wishing to update their older M48s.

While the basic M48 design is now over thirty years old, this last model of the Patton series is still able to hold its own against the majority of foreign tanks now in service throughout the world. During the next few years, however, the balance will shift against M48 due to new designs and features, but despite this it will be some time before the last operational M48 is consigned to the scrap heap.

# Development

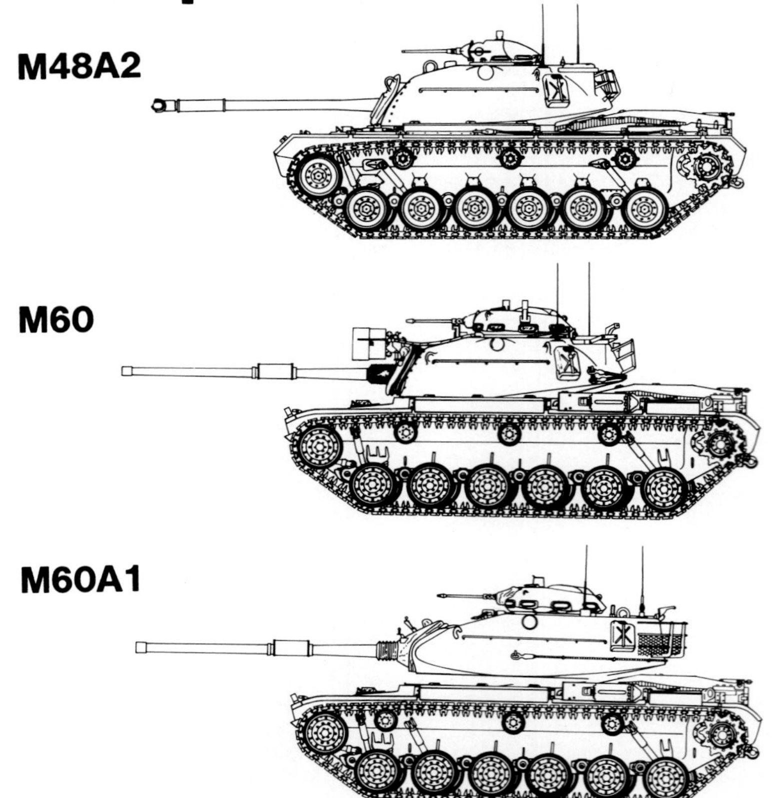


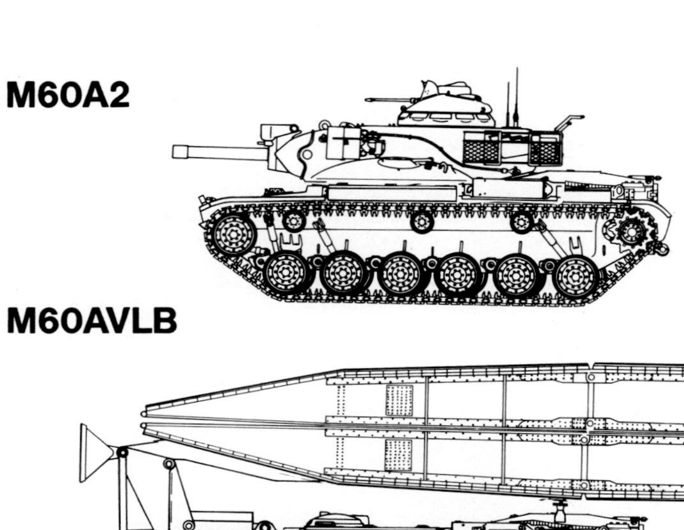


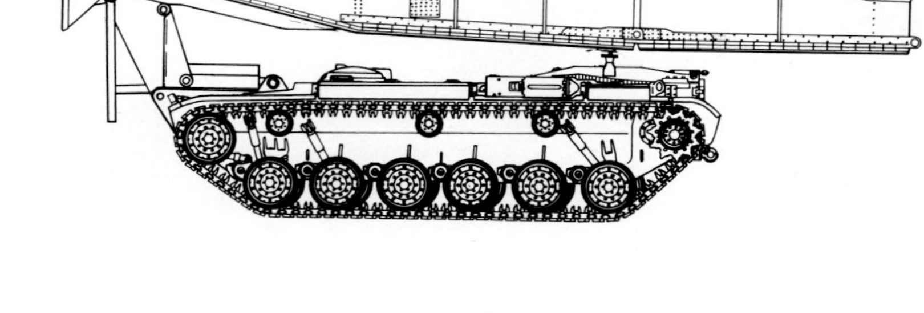

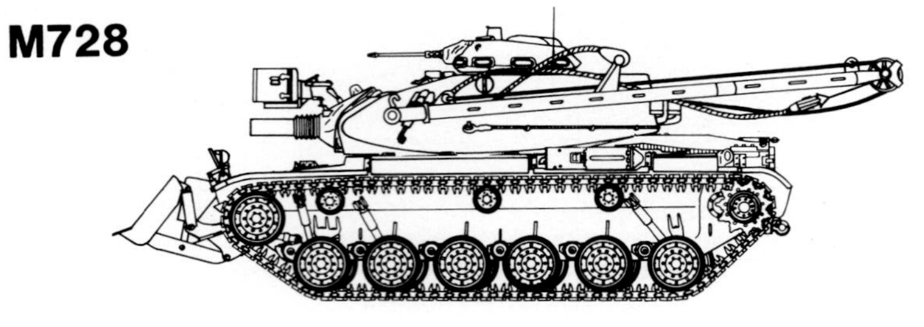


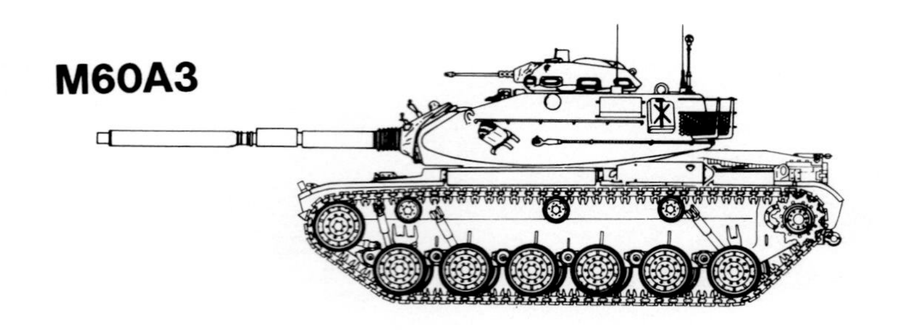

# M 60

By the mid-1950s, as the numerous problems which had beset the M-48 were being ironed out, detailed information on the new Russian T-54 was acquired by British intelligence and passed on to the Pentagon. This secret report showed that the new Soviet tank was very heavily armored and was armed with a 100MM gun which was superior to the 90MM gun arming the M-48. The US Army immediately began looking for ways to further update the basic M48 design and increase its firepower in order to counter the new Soviet tank.

Fortunately, the British had earlier realized the need for a heavier weapon to combat future generations of Russian armor and were already working on a new 105MM gun which was scheduled to be installed in their Centurions. The US Army procured one of these weapons and tested it in the T95 experimental tank with very impressive results. The British 105MM gun was found to be far superior to any gun currently available to the US Army, or to any gun undergoing tests. After redesigning the breech, the decision was made to use this weapon in the next series of American tanks. Under the designation M68 the British 105MM gun was fitted into the turret of an M48A2* and successfully tested. However, despite the ease with which the new gun was adapted to the M48 turret, the army wanted to further improve the basic Patton design before putting the up-gunned vehicle into production.

One of the major problems experienced with the M48 series had been the lack of adequate range of the M-48's gasoline engine so the army decided to switch over to a diesel engine for fuel economy, reduced fire hazard, and simplified logistics (most European armies either used the diesel, or favored the diesel). An AVDS-1790-2 powerplant was mated to a redesigned M48 hull which had the new 105MM gun fitted in the turret. This re-designed vehicle, designated the M60, was considerably more than just a re-engined and up-gunned M48. The most obvious external changes in the new vehicle were the replacement of the older M1 cupola with the new roomier M19 cupola, and the boat shaped front hull being replaced by a flat angular glacis plate which gave the new tank a more square shape. Other external features of the M60 included more rounded fenders and redesigned roadwheels. Internal modifications were made in order to accept the new engine and gun but there were no major changes in the general layout of the interior. The troublesome stereoscopic rangefinder was replaced with a more efficient coincidence rangefinder which was easier to operate and far more effective. Modifications were made to equipment which took advantage of technological advances, but these were basically updates of older systems.

The Army ordered the M60 into production in 1959, placing an order with Chrysler for an initial run of 180 vehicles which was followed by an additional order for 720 tanks. The first M60s entered service with US Army units during the fall of 1960, with most of the initial production vehicles being sent to Europe to off-set the Russian T-54 then coming into widespread service with Warsaw Pact armies. While it was an improvement over the M48, especially in armament, the M60 was regarded as somewhat of a stopgap measure. In particular the turret armor and design were criticized because the army felt that the newer Soviet ammunition could penetrate it. As a result the army decided to design a new turret even as it ordered additional M60s from Chrysler.

**For further information on this installation see the author's M48 PATTON IN ACTION published by Squadron Signal Publications.*

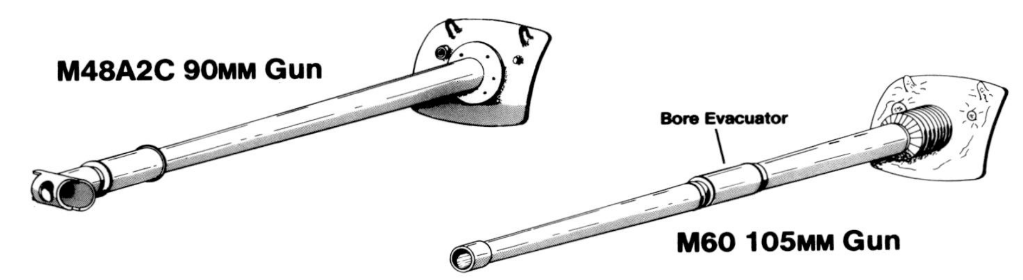


**(Right) The first pilot model of the M60 displays a remarkable similarity to the M48. The most notable differences are the 105MM gun, the M19 commander cupola, and new road wheels. Production vehicles would carry external shock absorbers. (US Army via Binder)**

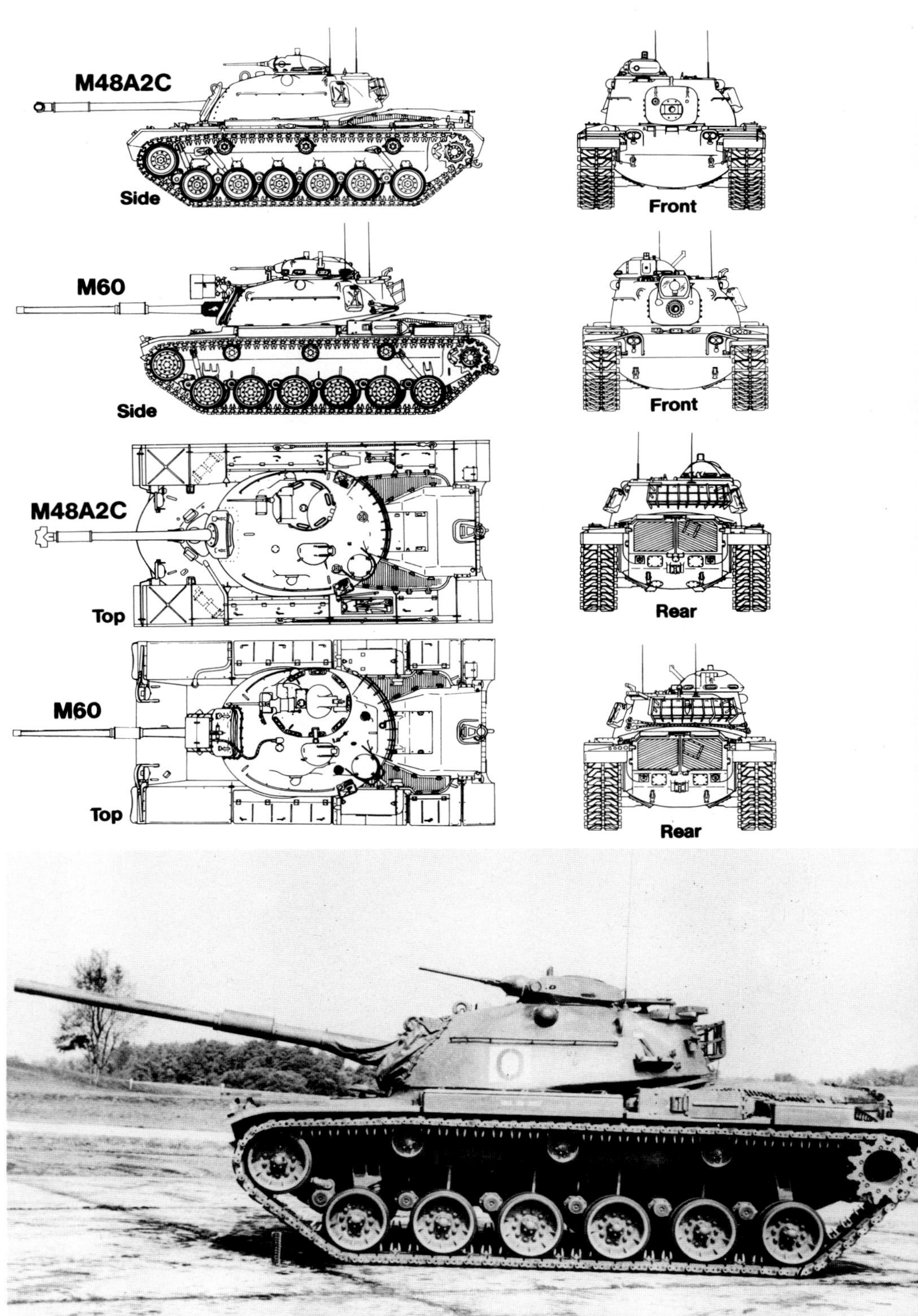

(Above) The M60 could be fitted with an M9 bulldozer blade kit, and could be closed up for deep water wading in a completely submerged condition. The wading tower atop the turret could have sections added or deleted as conditions demanded. (US Army via Binder)

(Above Left) A prominent change between the M48 and the M60, was a completely redesigned hull which incorporated an angular glacis plate instead of the curved boat-shaped forward hull of the M48. Note the curved fenders of the M60 and the distinctive shape of the headlight guards. (Chrysler via Binder)

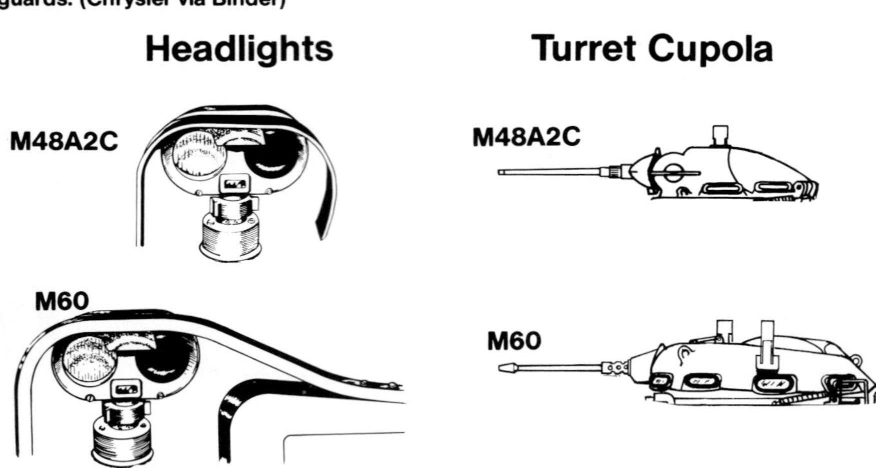

(Left) A completed turret on the Chrysler production line is being lowered onto the hull of an M60. The storage basket is built directly on the rear of the turret. (Chrysler via Binder)

(Above) Mines have always been a serious threat to armor and over the years various attempts have been made to alleviate the threat. This mine plow was developed for use on the M60 and could be fitted to all later variants of the series. It plowed up a furrow of earth in front of each track, including any mines, depositing it along side the moving tank. The mines were later disposed of by engineers. The chain and bar across the bow hung down to detonate any tilt rod mines which the plows miss. (US Army via Binder)

(Above Right) M60s were rushed to Europe to counter the threat posed by the new generation of Russian tanks. This M60 of the 2nd Battalion, 33rd Armored Regiment, 3rd Armored Division takes part in *Exercise Big Lift* during October of 1963 in Germany as part of the aggressor force. (US Army)

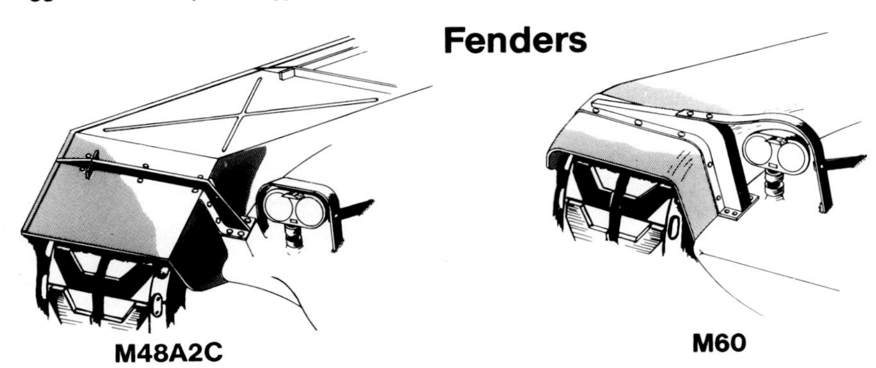


(Right) A tank is only as good as its crew and constant training is a must if the crew is to operate at peak efficiency. The crews of these M60s take part in a live firing exercise at Hohenfels, Germany. Each vehicle is equipped with a 2.2 kilowatt xenon searchlight. An abundance of crew gear is carried in the storage racks. (US Army via Binder)

(Above) During the early 1970s the Seventh Army began experimenting with multi-color camouflage patterns in an attempt to better conceal its tanks in the field. The patterns varied from vehicle to vehicle. (Seventh Army via Binder)

## Road Wheels

## Tension Wheels

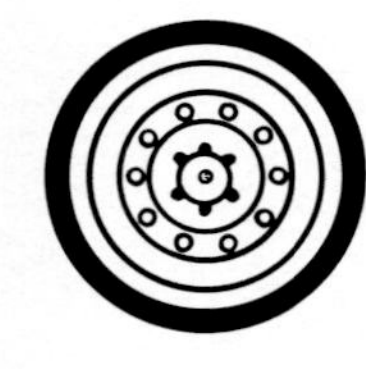

M48A2C

M60

M48A2C

M60

(Above Left) The original M60s sent to Europe were painted Olive Drab since the army did not feel additional camouflage paint would provide any real advantage in North Western Europe. In winter, however, Olive Drab stood out like a sore thumb and tank crews often gave their vehicles a rough coat of White paint to blend in with the snow covered terrain. This M60 of the 1st Battalion, 64th Armored Regiment, 3rd Infantry Division crosses the Man River during a winter exercise. During such close driving conditions a ground guide was usually needed to assist the driver whose view was restricted. (US Army via Binder)

(Left) Painted Sand, Red Brown, Green, and Black this M60 of the 3rd Infantry Division waits in ambush under a bridge near the town of Gunzenhausen, West Germany during OPERATION CERTAIN CHANGE. (Seventh Army via Binder)

# M60 105mm Gun Tank

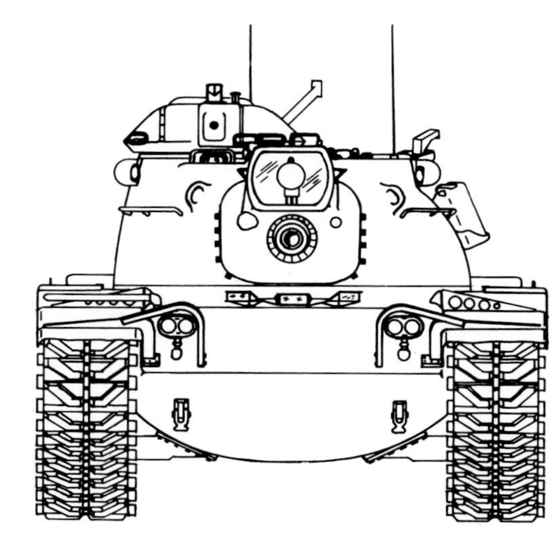

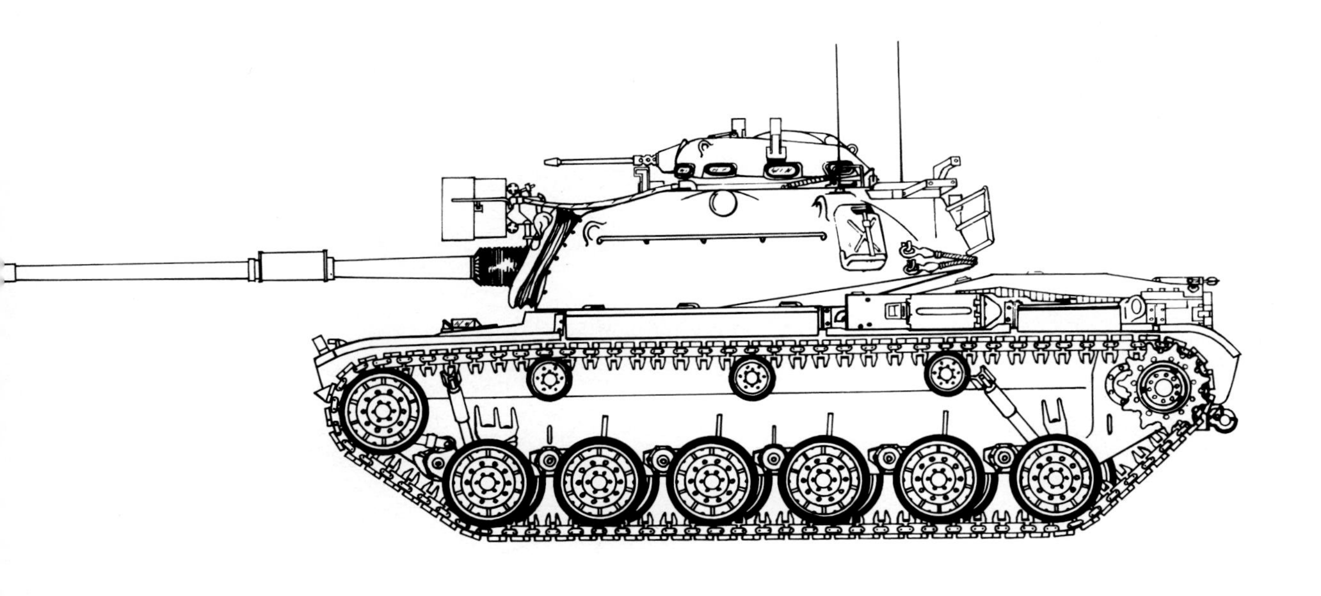

## SPECIFICATIONS

### 105MMGun Tank M60

| | |
|---|---|
| **Crew** | 4 men |
| **Length (gun forward)** | 366.5 inches |
| **Width** | 143.0 inches |
| **Height (over cupola periscope)** | 126.5 inches |
| **Ground Clearance** | 15.3 inches |
| **Combat Weight** | 102,000 pounds |
| **Armament** | |
| **Main** | 105MM Gun |
| **Secondary** | |
| **Cupola Mount on Turret** | 50 Caliber Machine Gun |
| **Co-Axial** | 7.62MM Machine Gun |
| **Engine** | AVDS-1790-2 |
| **Horsepower** | 750 HP |
| **Maximum Speed** | 30 MPH |
| **Maximum Range** | 300 Miles |

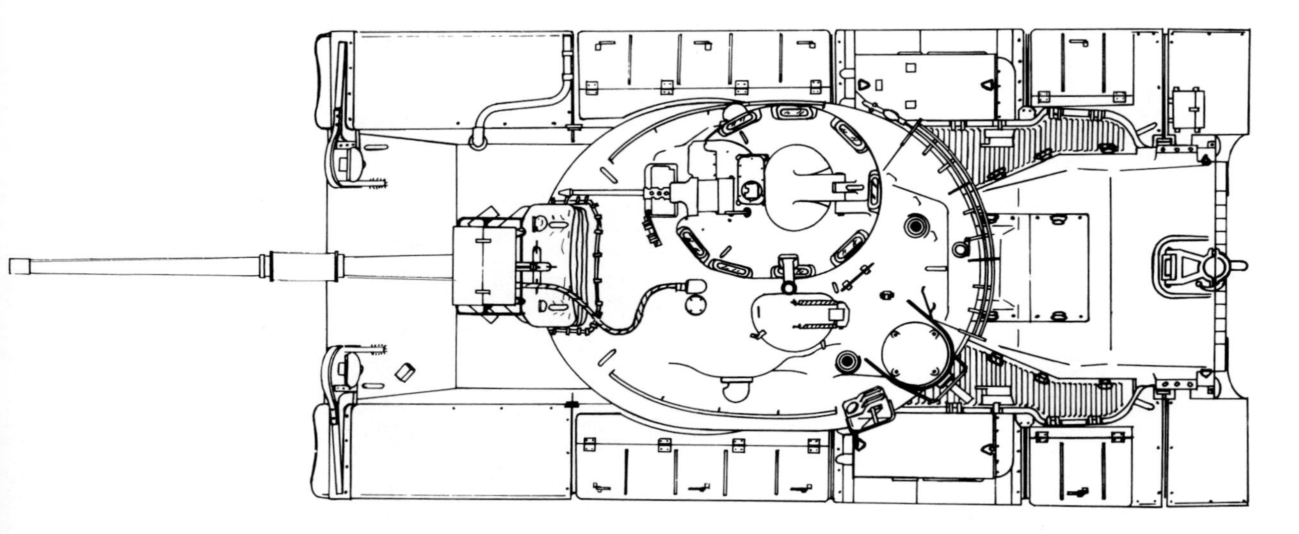

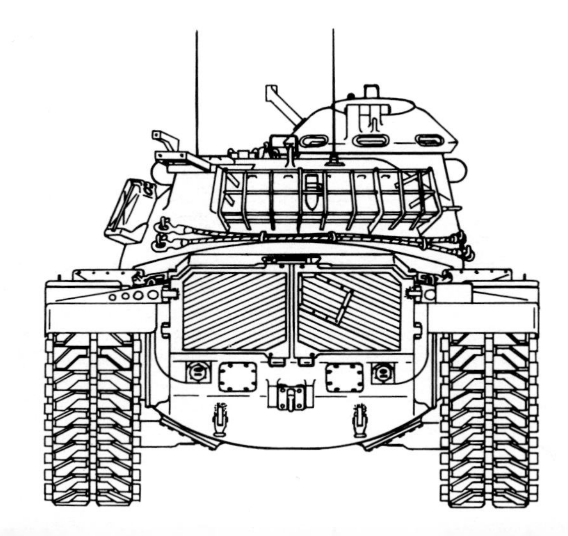

# M60A1

While the M60 provided the US Army with an excellent interim tank with which to counter the Russian T-54, even while it was under development the army decided to redesign and increase the armored thickness of the turret. The new turret design featured an elongated nose which provided better ballistic protection than the turtlebacked shaped turret of the M48. In addition it was wider and had a pronounced rear bustle, making more interior space available to the crew. The new turret design, necessitated some rearranging of equipment, but no drastic changes in the overall layout. Other than the new turret design, little was done to the basic M60 chassis excepting minor changes in hull fittings. The new variant, under the designation M60A1, was able to be placed in production relatively quickly, and without serious problems.

The first M60A1s were issued to regular army units during the spring of 1962, less than two years after the first M60s were placed in service. Following introduction of the M60A1 into American service, it was supplied to US allies, including Austria, Iran, Israel, Jordan, and Italy. In addition the Italians were granted a license to produce the vehicle, with the firm OTO Melara eventually producing 200 M60A1s for the Italian Army.

## M60A1(AOS)

With the M60A1 the US Army had a tank with good armor protection, a hard hitting long range gun, and a robust and reliable performance. Throughout the 1960s the M60A1 gradually became the main battle tank of the US armored forces. However, by the end of the decade the army realized that to keep pace with Russian armored advances, the M60A1 would have to be put through a modernization program. The first step in the up-dating process began in 1971 when an Add-On Stabilizer (AOS) system was fitted to the gun, allowing it to remain on target while the tank was moving. Other improvements included replacing the side loading airfilters with top loading air filters, and replacing the T 142 track with octagonal track pads. Tanks so modified received the designation M60A1(AOS).

## M60A1(RISE)

The next stage in upgrading the M60A1 was begun in 1974 when the AVDS-1750-2D diesel engine was replaced by the improved AVDS-1790-2D diesel engine as part of a reliability improvement of selected quipment (RISE) program. The AVDS-1790-2D powerplant had been tested and found to have a much longer operational service life. When the 300 ampere electrical system was replaced with a 650 ampere oil-cooled alternator, a solid state regulator, and a quick disconnect wiring harness were mated to the RISE engine, the powerplant was designated the AVDS-1790-2C. The M60A1(AOS) fitted with the AVDS-1790-2C engine was given the designation M60A1(RISE), and began coming into service during 1975.

## M60A1(RISE/PASSIVE)

In 1977 the last major series of improvements for the M60A1 variant began with the addition of a deep water fording system and a passive night sight. M60A1(RISE) tanks with this configuration were designated the M60A1(RISE/PASSIVE). This new passive night system, unlike previous infra-red systems, did not require illumination from the tank. Instead, the tank commander/gunner's sight (M35E1), and the drivers sight (AN/VVS-2) used an image intensification which amplified ambient starlight. To guard against a systems failure or the lack of sufficient starlight, the infrared viewers and AN/VSS-2A searchlight were retained.

In addition to these modification programs a number of other changes were made. During the 1973 Yom Kippur, a number of M60s were lost after taking hits under the chin or the turret ring. An armored chin fillet was added and the turret ring was increased in thickness. The hydraulic fluid in the turret traverse system was found to have a low flashpoint which resulted in a number of tanks being lost to fire after taking relatively minor battle damage. A new hydraulic fluid with a higher flashpoint was substituted.

In 1978 another modification program began, however, because of the number of changes involved the army decided to redesignate these modified vehicles to M60A3s. Some of these changes are currently being retrofitted to existing M60A1's but without a change in designation. Whether or not the entire M60A1 fleet will be retro-fitted with all the changes of the M60A3 model is doubtful since the new M1 Abrams has been procured as the US Army's main battle tank.

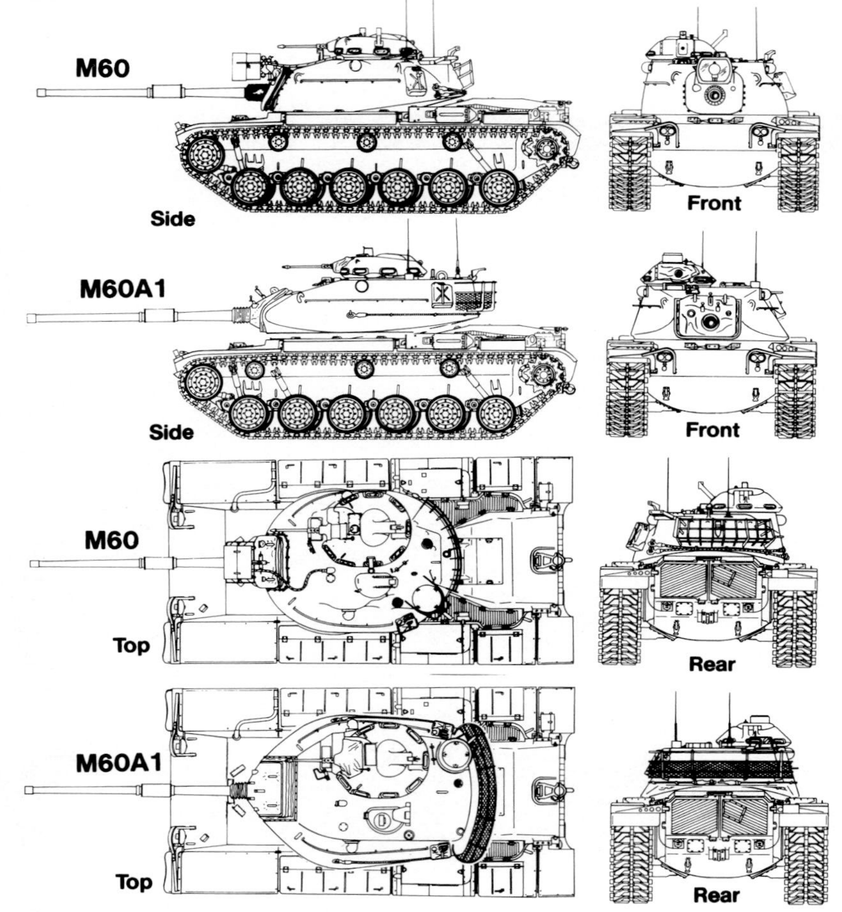


(Below) The Number 3 pilot model of the M60A1 which was shipped to Fort Knox in July of 1961. The most noticeable difference between the M60 and the M60A1 was the new elongated turret which provided much better ballistic protection than the older turret based on an M48 turret. (US Army via Binder)

(Above) A Chrysler employee works on the inside of a turret on the Detroit assembly line. The depression at the bottom rear of the turret provided added clearance for the driver when the turret is traversed to the rear. (Chrysler via Binder)

(Above Right) Welders attach suspension components to a hull. A great deal of work must be done to a tank before it is ready for acceptance by the army. (Chrysler via Binder)

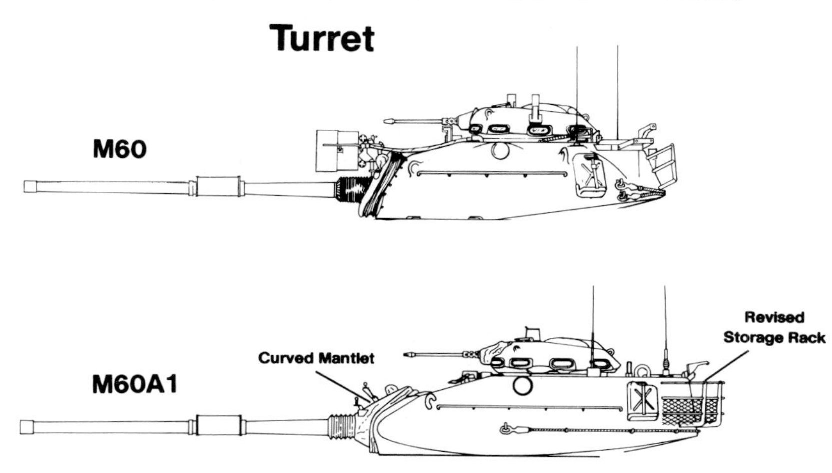

(Right) Turrets move along the assembly line before being mated to a hull. The turrets on the left still need their M19 cupolas fitted. (US Army via Binder)

(Above) The end result of all this labor - a nearly completed M60A1. Workers lower the turret onto the hull before all the minor fittings have been added. Once these fittings have been added the tank will be subject to a series of tests prior to the army accepting it for service. (Chrysler via Binder)

## Storage Rack

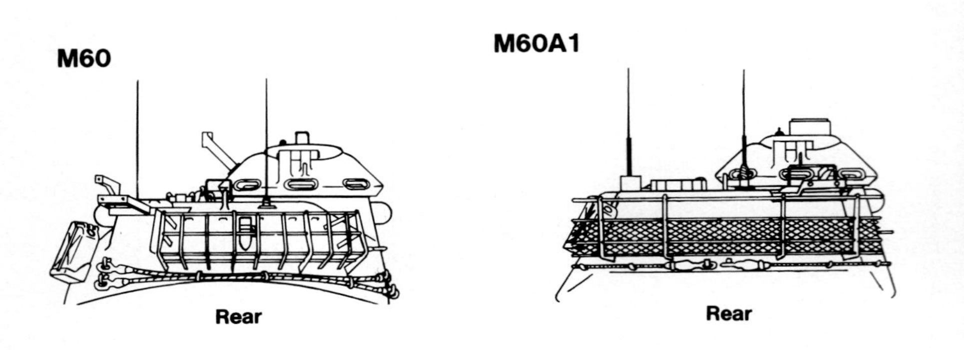

(Above) Not all turrets were mated to tank hulls, this turret is being used as a trainer for armored troops at the Fort Knox Armor School. The turret side has been cut away to facilitate instruction of the interior. (US Army/Fort Knox via Binder)

(Below) A partial view of the drivers compartment. The tubes on either side of the seat are storage racks for 105MM ammunition. Directly behind the drivers seat are the tank's batteries. (Binder)

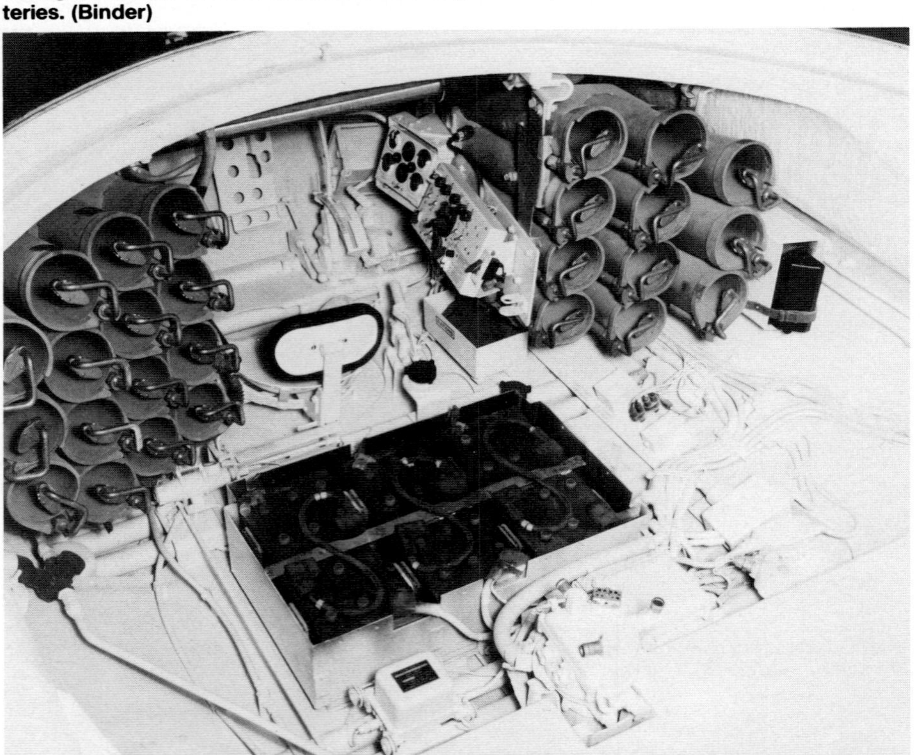

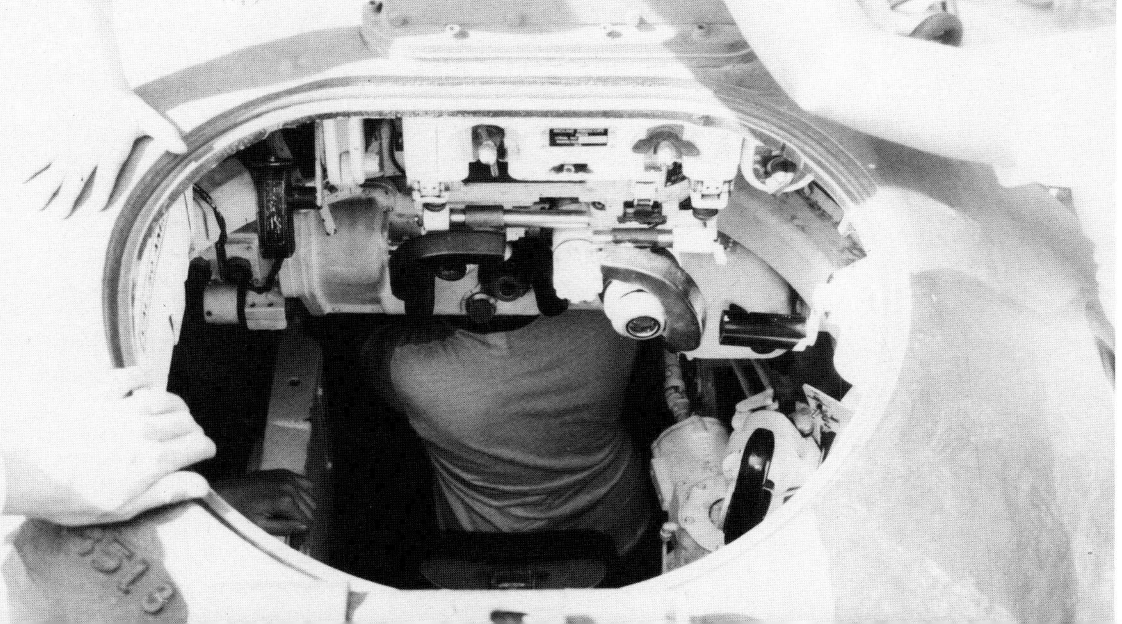

**(Above) The major types of ammunition carried by the M60A1: (From left to right) HEP - high explosive plastic, WPO - white phosphorous, APDS - armor piercing discarding sabot, and HEAT - high explosive anti-tank. (Binder)**

**(Above Left) The M60A1 can be fitted with an M9 Bulldozer blade kit. Normally one tank per company was fitted with the kit to help dig positions for the rest of the tanks. This particular tanks is from the 1st Infantry Division during OPERATION REFORGER V during the fall of 1973. (US Army via Binder)**

**(Left) The periscope and rangefinder are located at the top center of the commander's cupola. On the right is the cupola traverse knob and on the left is the machinegun elevation wheel. (Binder)**

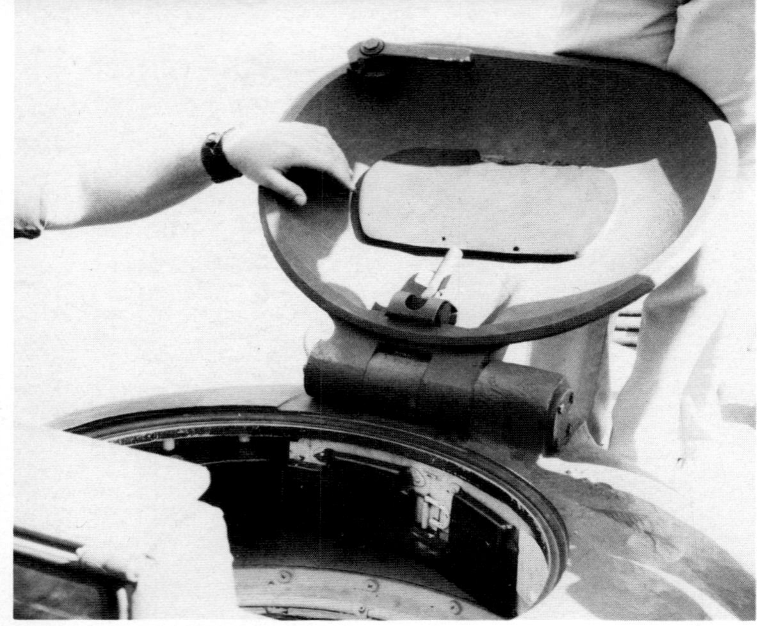

(Above) The protective pads can be seen in the center of the cupola hatch and just inside of the cupola. The cover of the periscope has a spring to snap it down. (Binder)

(Above Left) An early production M60A1 prior to the modifications instituted to upgrade it in order to bring it up to the standards of tanks coming into service with the Warsaw Pact nations. This particular vehicle is from the 32nd Armor. (Teledyne Continental Motors via Binder)

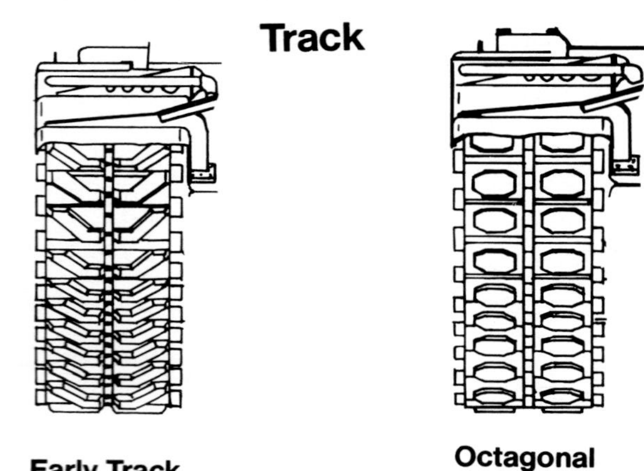

(Left) Some of the modifications carried out to upgrade the M60 series were the M142 tracks with their octagonal rubber blocks and the chin armor which can be seen behind the drivers head. In addition, new top loading air cleaners were also part of the upgraded. (3rd Armor via Binder)

**(Above) This tank, although fitted with the new chin armor, still has the old style track. Mud has been used to cover up the white vehicle markings to make the tank less conspicuous. (US Army via Binder)**

**(Above) This M60A1 from the 3rd Battalion, 32nd Armor, 3rd Armored Division has been fitted with the storage bins from an old M47 Patton on the back of the storage rack. The tank is in a four color scheme consisting of Sand, Red Brown, Green and Black. This scheme was peculiar to elements of the Seventh Army during the mid-1970s. (Binder)**

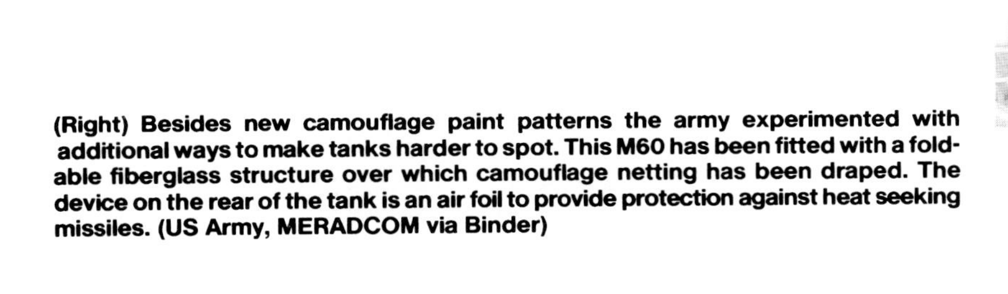

**(Right) Besides new camouflage paint patterns the army experimented with additional ways to make tanks harder to spot. This M60 has been fitted with a foldable fiberglass structure over which camouflage netting has been draped. The device on the rear of the tank is an air foil to provide protection against heat seeking missiles. (US Army, MERADCOM via Binder)**

(Above) A Marine Corps M60A1 comes ashore during a landing in the Caribbean. During the invasion of Grenada five Marine M60s were used to support the Marine and Army troops. (USMC)

(Above Left) Track changes are one of the least desirable jobs that tankers performed. The shape of the new M142 track is very distinctive. (3rd Armor via Binder)

## Xenon Searchlight

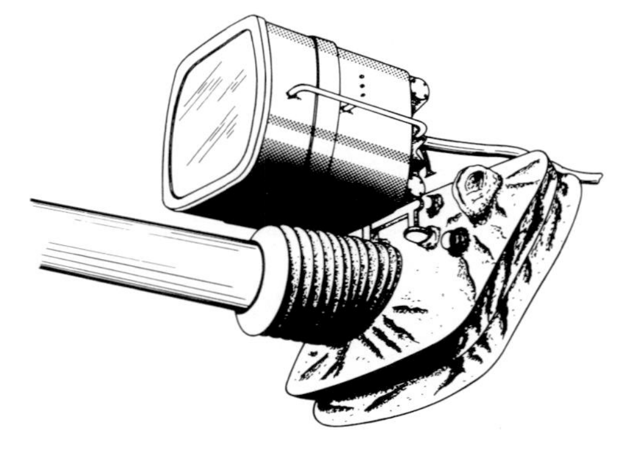

(Left) An M60A1 in a four color camouflage scheme is being loaded aboard a transport during REFORGER in 1977. The vehicle still carries the older side loading air cleaners. (US Army via Binder)

(Above) An M60A1 (Rise/Passive) under cover in a German forest during summer exercises. The commander has rigged a sunscreen using the antennas as a framework . (3rd Armor via Binder)

(Above Right) This M60A1 has been fitted with a smoke generator and 55 gallon fuel drums to test the feasibility of the new smoke system. The smoke system is similar in makeup to the one used by the Russians on their tanks. (US Army via Binder)

(Right) The smoke generating system undergoing tests at Fort Knox. Eventually a modified smoke system was standardized on the M60A3 variant which began to appear during the latter part of the 1970s. (US Army via Binder)

# M60A1 105mm Gun Tank

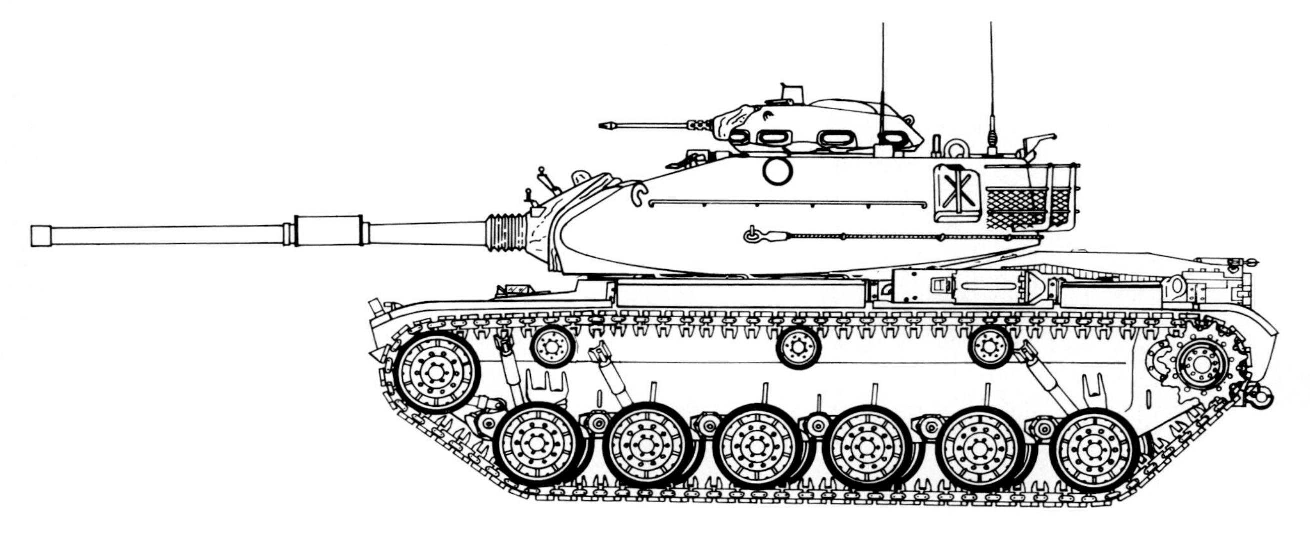

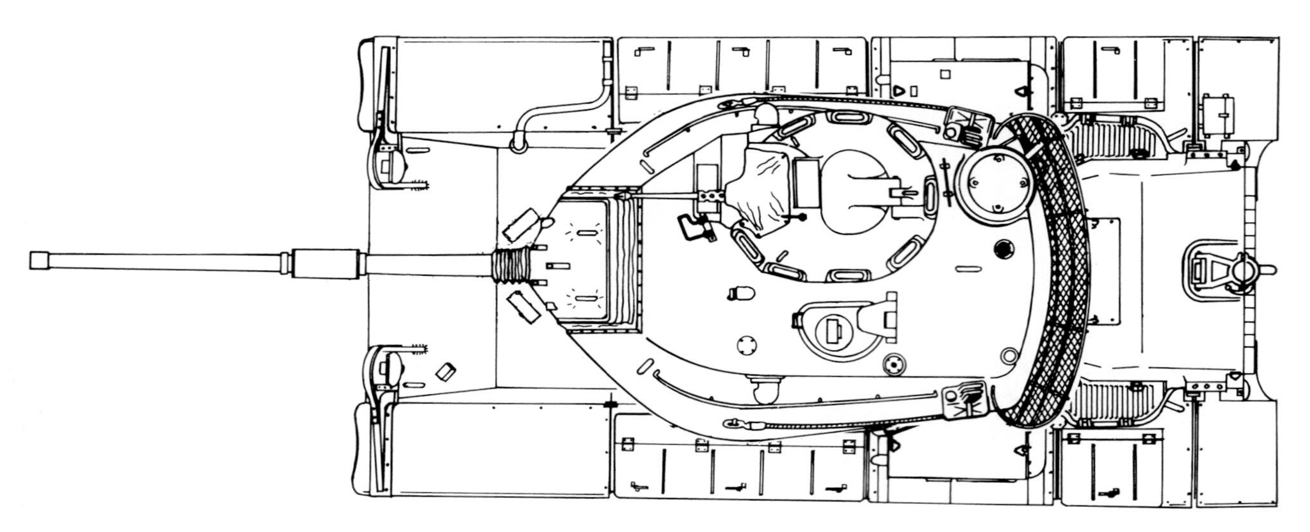

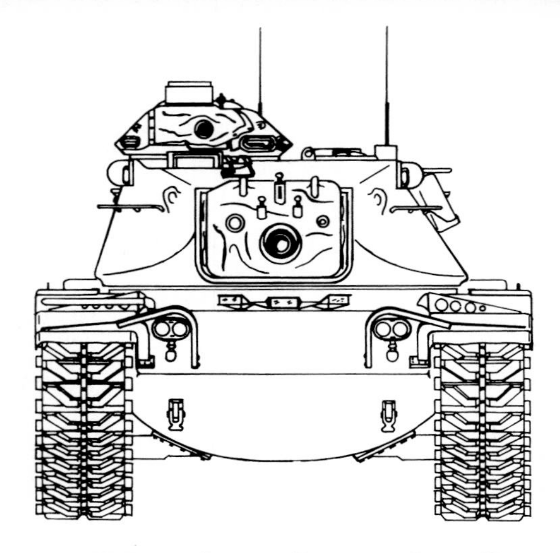

## SPECIFICATIONS

### 105MM Gun Tank M60A1

| | |
|---|---|
| **Crew** | 4 men |
| **Length (gun forward)** | 371.5 inches |
| **Width** | 143.0 inches |
| **Height (over cupola periscope)** | 128.5 inches |
| **Ground Clearance** | 15.3 inches |
| **Combat Weight** | 105,000 pounds |
| **Armament** | |
| **Main** | 105MM Gun |
| **Secondary** | |
| **Cupola Mount on Turret** | 50 Caliber Machine Gun |
| **Co-Axial** | 7.62MM Machine Gun |
| **Engine** | AVDS-1790-2A |
| **Horsepower** | 750 HP |
| **Maximum Speed** | 30 MPH |
| **Maximum Range** | 300 Miles |

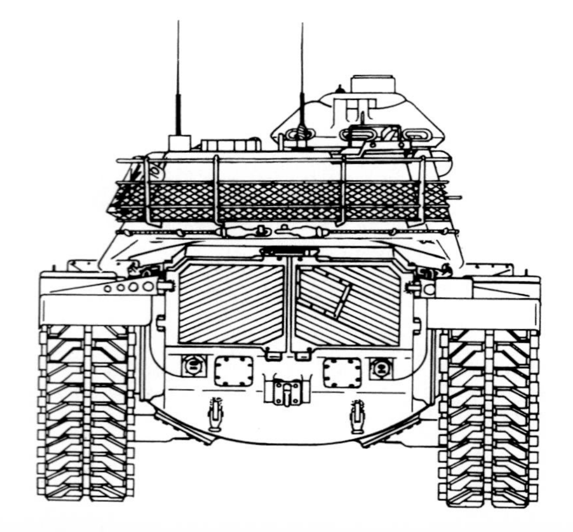

# M60A2

The most unique variant in the M60 series was the M60A2, designed to provide the army with a long range missile firing tank to supplement the firepower of the M60 and M60A1 then in service in Europe. During the mid 1960s, the army decided to mount the newly developed Shillelagh missile system on an M60 chassis. This system, originally developed for use in the new M551 Sheridan Armored Reconnaissance/Airborne Assault Vehicle (AR/AAV), had superior range and far greater accuracy than the standard 105MM gun of the M60 series. The initial program, designated the XM 66, tested four basic turret designs, labeled A thru D, to see which would prove most effective. A and B turrets were totally new designs incorporating a smaller profile, better ballistic shape, improved armored protection, and a 20MM cannon mounted at the rear of turret. The C turret was a more conventional turret with a 20MM cannon mounted in a cupola, while the D turret was basically a modified M60A1 turret.

The Army opted for the B turret because it felt that the reduction in turret weight would significantly increase the tanks mobility. The entire program was viewed as a *low risk* venture since it was the mating of an existing systems with a new turret. Unfortunately, this proved erroneous. Two test vehicles, under the designation M60A1E1, were delivered to the army between November of 1965 and February of 1966. During the test program numerous problems arose with both the Shillelagh armament and the new turret. The XM81E13 152MM gun-launcher (eventually redesignated the M162), used combustible cased ammunition which caused a number of problems. The ammunition rounds could not take rough handling and were found to be highly susceptible to humidity. In addition, there were flash backs and premature detonation of rounds in the breech caused by remnants of the previous rounds which had not been consumed. When firing conventional ammunition the recoil of the gun jarred the light turret, throwing the fire control system off its target. Eventually these problems were solved, but it took time, and delayed the operational debut of the new tank. To solve the shell casing residue problem the army added a new closed breech scavenging system (CBSS) which cleared the gun by blowing three blasts of compress air through it. Two compressors and storage bottles were installed in the lower rear hull resulting in a bulge below the rear doors of the engine compartment. Recoil of the gun automatically initiated the scavenging sequence prior to breech opening. With the addition of the CBSS there was no need for the bore evacuator and it was eliminated from later production vehicles. In order to eliminate the jarring effect on the fire control system its turret mounts were reinforced.

Even as these problems arose the army placed orders for the new tank, funds being provided in 1966 for 243 of the new M60A1E1 turrets to be fitted on existing M60 chassis, with plans to procure 300 complete tanks for 1967 under the designation M60A2. Eventually the army received 540 of the Shillelagh armed M60 tanks (although Chrysler records only indicate that they completed only 526). Due, however, to the problems with the test models, production did not begin until 1973 and lasted only into 1975.

While all this was going on Chrysler designed a new turret of their own as an alternative. The Chrysler turret carried a modified version of the gun fitted to the experimental MBT70 then undergoing tests. The gun, designated the XM150, was long barrelled and had the capacity to fire both the already available Shillelagh rounds as well as kinetic energy armor piercing rounds. The turret itself was composed of rolled armor sloped at high angles. The same fire control system used on the M60A2 was installed in the Chrysler turret but a semi-automatic loader for conventional rounds was added to the gun which did not interfere with the use of the missile rounds. In addition more rounds (57 versus 46) could be carried. All this was achieved for a modest weight increase of 1800 pounds. Another variation of this turret was armed with a 120MM Delta gun, a hypervelocity weapon developed from an earlier 90MM and 105MM smoothbore gun. This gun added about 1000 pounds to the basic design. Unfortunately this vehicle, called simply the 'K' tank, was abandoned despite its potential due to a lack of funds which were being used on the war in Vietnam. A powerful new armored weapon might have been added to the Army's arsenal.

By the mid-1970's the M60A2 was ready for issue to armored units. Six armored battalions in Europe were equipped with the new tank but it never became popular with its crews, due to the com plexity of the gun and the need for highly skilled maintenance. Due to this complexity troops in the field dubbed it with the *very* unofficial nickname *Starship*, and it was the subject of a great deal of unprintable comments by its crews. As a result of the strides made in high performance kinetic ammunition and fire control systems the advantage of the M60A2's missile armament became minimal and the army moved to scrap the troublesome turrets. Many of the hulls were later used for special purpose vehicles such as bridgelayers.

## M60A2 Turret

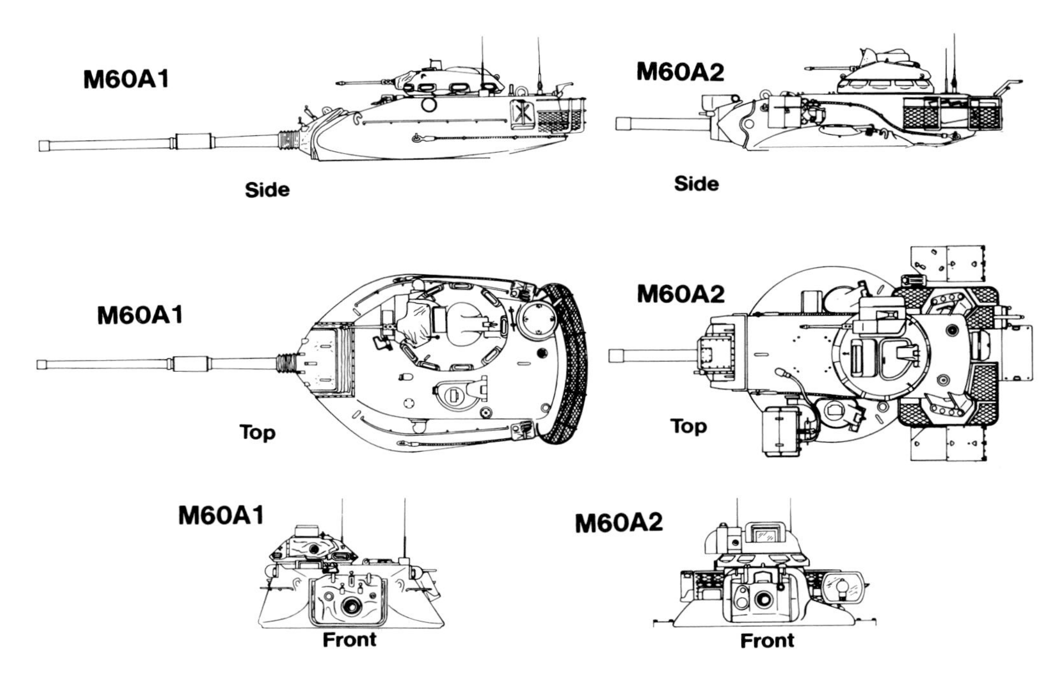


**(Below) The A model turret was a compact design. Both the A and B model turrets were similar in design and concept. Eventually the B turret was selected for continued development. (Icks via Binder)**

(Above) The D turret was designed around the M60A1 production turret, but with a shortened turret bustle for the D model. (Icks via Binder)

(Above Left) The hull used to test the D turret was eventually fitted with a mockup of the turret that finally went into production. The early production turret had a bore evacuator fitted to the gun tube. (US Army via Binder)

## M60A2 (Early)

(Left) The M60A1E1 Advanced Production Engineering (APE) pilot number 1 is seen on the grounds of Fort Knox during the spring of 1967. The lines of the production turret have been squared off in comparison to the mock up, the smoke grenade launchers have been deleted and changes were made in the turret storage racks. (Chrysler via Binder)

(Above) The compact shape of the turret as well as the location of the various hatches for the crew can be seen from above. The smoke grenade launchers have been repositioned to the rear of the turret in the midst of the storage racks. (Icks via Binder)

(Above Right) This M60A1E2 is fitted with a later version of the 152MM gun-launcher without a bore evacuator on the gun tube. Due to problems with smoldering residue a closed breech scavenger system was fitted to later models to clean out the barrel with compressed air. Tanks fitted with this system had a bulge on the lower rear of the hull for the necessary equipment. (Icks via Binder)

## M60A2 (Late)

(Right) Chrysler developed the 'K' turret as an alternative to the turret selected by the Army. The prototype of the 'K' turret is being lowered onto a modified M60 hull. While the 'K' tank offered a number of benefits over the M60A2, funds for its development were unavailable due to other priority projects and the war in Vietnam. (Chrysler via Binder)

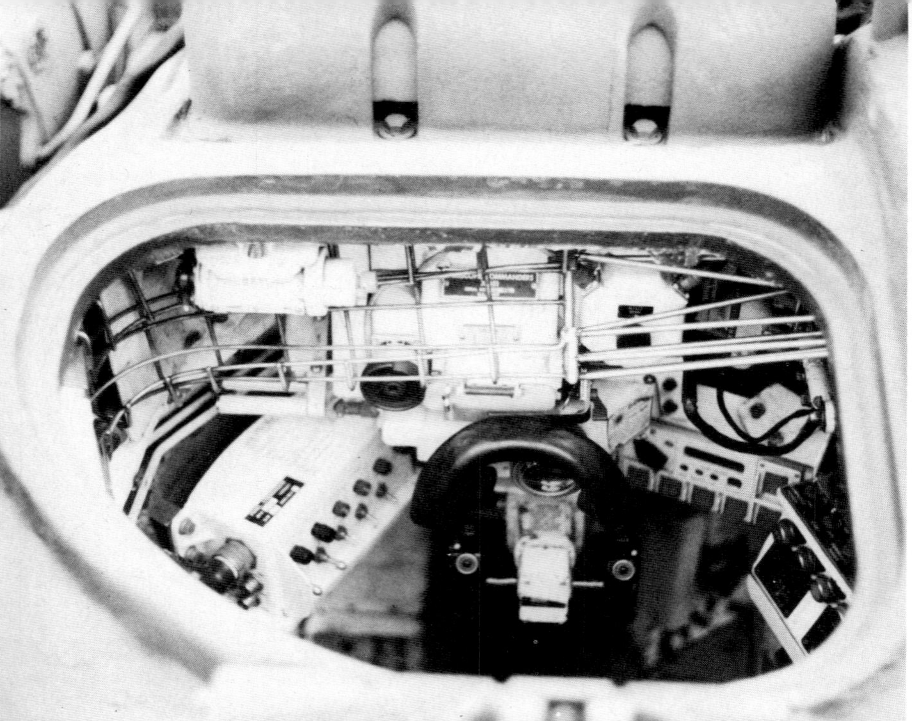

(Above) Some of the controls can be seen through the tank commanders cupola. The sighting device is under the inverted U shaped piece of Black rubber. (Binder)

(Above Left) The xenon searchlight mounted on the M60A1E2 helped extend the accurate range of the tank's missile during darkness when fitted with a pink filter, but even so its range was severely restricted at night. Under night conditions conventional ammunition was normally used. (Icks via Binder)

## 152MM Gun-Launcher

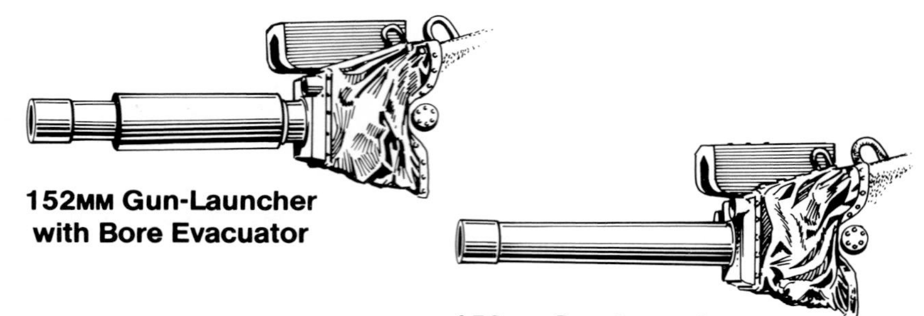

(Left) The breech, gun mount, and coaxial machine gun. The various hoses atop the breech mechanism are for the closed breech scavenger system (CBSS) mounted in production vehicles. (US Army via Binder)

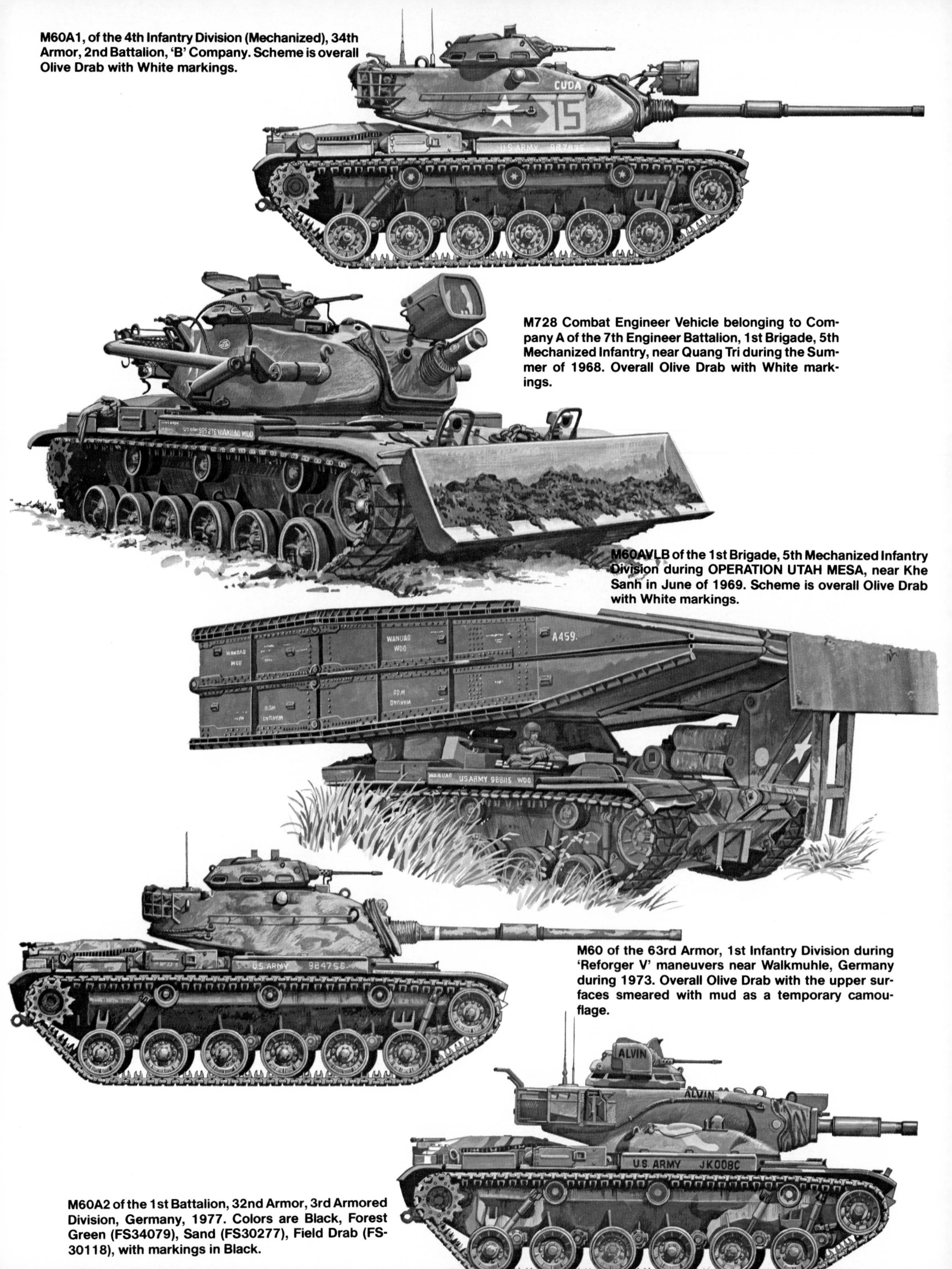


M60A1, of the 4th Infantry Division (Mechanized), 34th Armor, 2nd Battalion, 'B' Company. Scheme is overall Olive Drab with White markings.

M728 Combat Engineer Vehicle belonging to Company A of the 7th Engineer Battalion, 1st Brigade, 5th Mechanized Infantry, near Quang Tri during the Summer of 1968. Overall Olive Drab with White markings.

M60AVLB of the 1st Brigade, 5th Mechanized Infantry Division during OPERATION UTAH MESA, near Khe Sanh in June of 1969. Scheme is overall Olive Drab with White markings.

M60 of the 63rd Armor, 1st Infantry Division during 'Reforger V' maneuvers near Walkmuhle, Germany during 1973. Overall Olive Drab with the upper surfaces smeared with mud as a temporary camouflage.

M60A2 of the 1st Battalion, 32nd Armor, 3rd Armored Division, Germany, 1977. Colors are Black, Forest Green (FS34079), Sand (FS30277), Field Drab (FS-30118), with markings in Black.

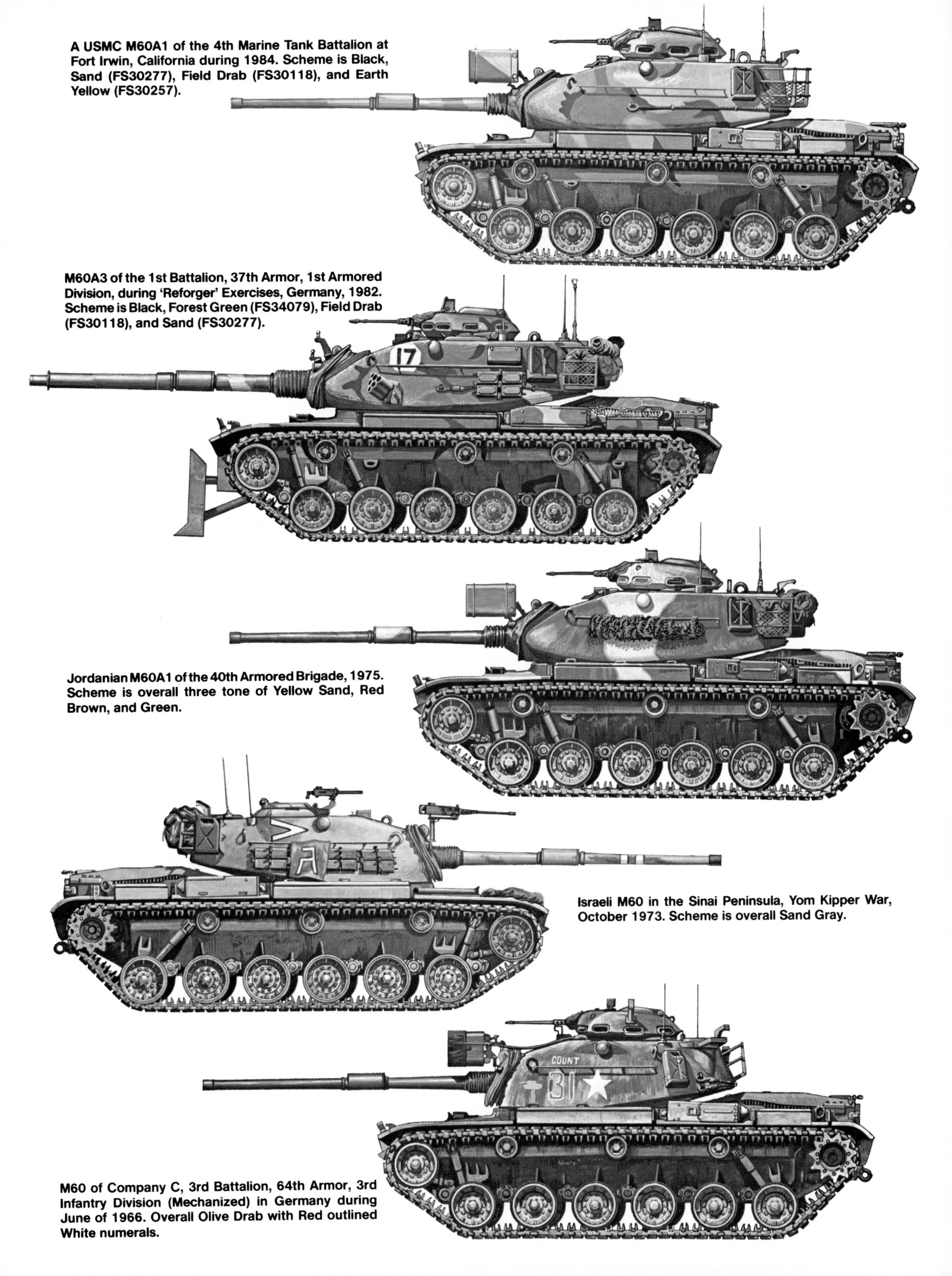


A USMC M60A1 of the 4th Marine Tank Battalion at Fort Irwin, California during 1984. Scheme is Black, Sand (FS30277), Field Drab (FS30118), and Earth Yellow (FS30257).

M60A3 of the 1st Battalion, 37th Armor, 1st Armored Division, during 'Reforger' Exercises, Germany, 1982. Scheme is Black, Forest Green (FS34079), Field Drab (FS30118), and Sand (FS30277).

Jordanian M60A1 of the 40th Armored Brigade, 1975. Scheme is overall three tone of Yellow Sand, Red Brown, and Green.

Israeli M60 in the Sinai Peninsula, Yom Kipper War, October 1973. Scheme is overall Sand Gray.

M60 of Company C, 3rd Battalion, 64th Armor, 3rd Infantry Division (Mechanized) in Germany during June of 1966. Overall Olive Drab with Red outlined White numerals.

(Above) The M60A1E2 pilot without the closed breech scavenger system while the M60A2 (Above Right) has the closed breech scavenger system as evidenced by the bulge in the lower rear hull to house the compressor and bottles. (Icks via Binder, 3rd Armored Division via Binder)

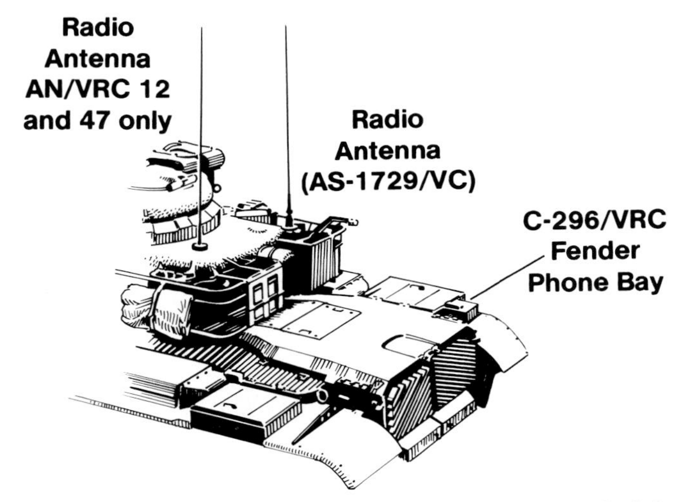

The rear of the xenon searchlight and its mount. The mounting arm moved, allowing the light to remain aligned with the gun. When not in use the light could be carried in a travel rack located on the rear turret storage rack. (Binder)

(Above) Perhaps the whole story of the M60A2 can be summed up in the legend on the barrel of this tank — 'COMPUTERIZED AGONY'. The tank was unpopular with its crews who felt it needed a great deal of skilled maintenance to handle the complex missile firing system, and once the higher performance kinetic energy weapons became available to upgrade the basic M60 armament the M60A2 was quickly phased out of service. (US Army, Fort Knox, via Binder)

(Below) A trio of M60A2s take part in missile firing training in Germany during the initial deployment of the M60A2. These tanks are from the 1st Battalion, 37th Armored, of the 1st Armored Division. Conditions such as these cut down on the range of the missile because of sighting difficulties, often negating the overall advantage of the missile system. (1st Armored via Binder)

(Above) An M60A2 is offloaded from a flatcar. This M60A2 was one of first received by the 3rd Armored Division and was assigned to the 1st Battalion of the 2nd Armor. (3rd Armored Division via Binder)

## MGM51C Shillelagh Missile

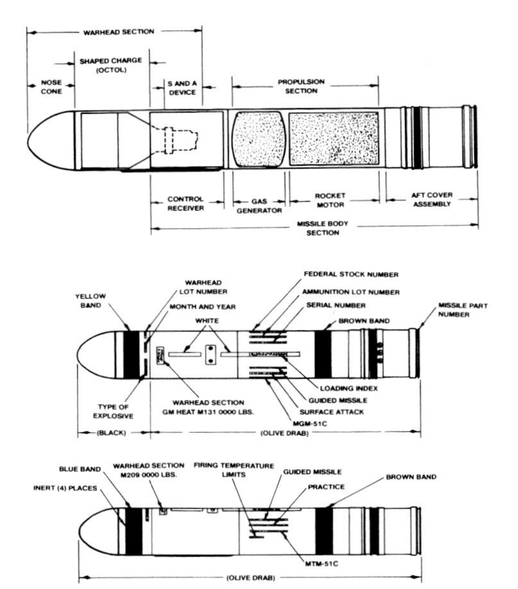


## Conventional Ammunition For 152MM Gun-Launcher

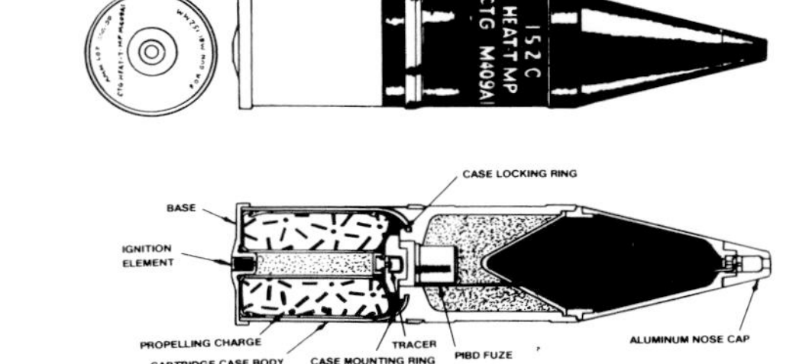

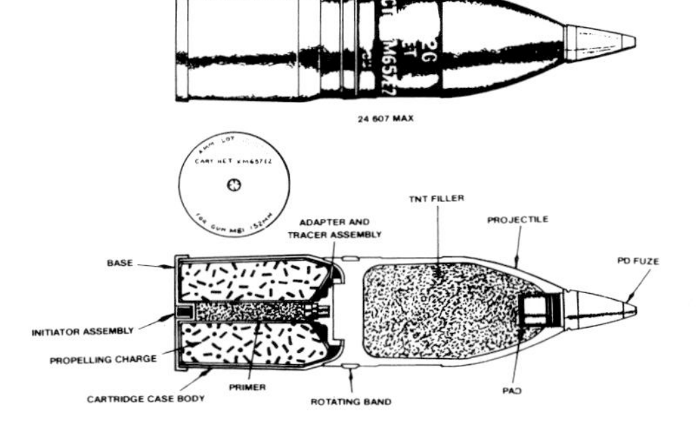


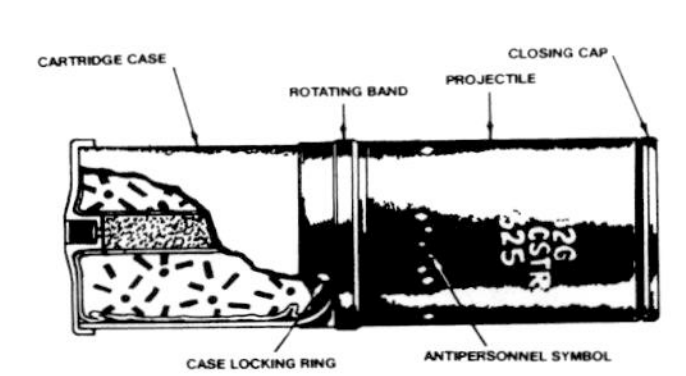

(Right) This M60A2 of the 3rd Armored Division passes through a small German town during a training exercise with its driving lights on. The weapon carried by the loader is the old M3 'grease gun' which is still the standard weapon carried by tank crewmen. (3rd Armored Division via Binder)

(Above) The new MERDC color scheme is seen on an M60A2 during the mid-1970s. MERDC schemes were adopted by the army to provide a series of four color disruptive patterns which could easily be altered to fit terrain or climate by changing one or two of the colors. In addition the various code markings and stars were painted in Black to reduce visibility at long range by enemy observers. (3rd Armored via Binder)

(Above Right) *BAM-BAM*, from the 32nd Armor, 3rd Armored Division, moves through a German forest during war games. Under such conditions the missile's range was severely limited due to obstructions and reduced visibility. (3rd Armored Division via Binder)

(Right) Due to its complexity the M60A2 was rather quickly withdrawn from service with the recycled chassis being used for special purpose vehicles. This M60A2 from the 1st Armored Division is moving down a dusty Texas road at Fort Hood in 1978. The six tank battalions which were equipped with the M60A2 variant were eventually switched over to the M60A1 as the A2s were withdrawn. (1st Armored Division via Binder)

# M60A2 152mm Gun Tank

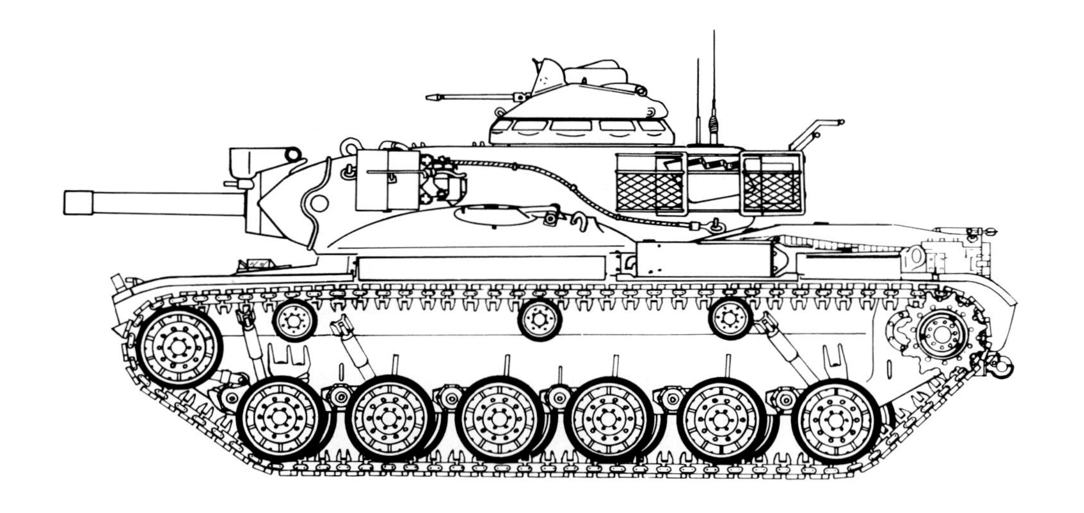

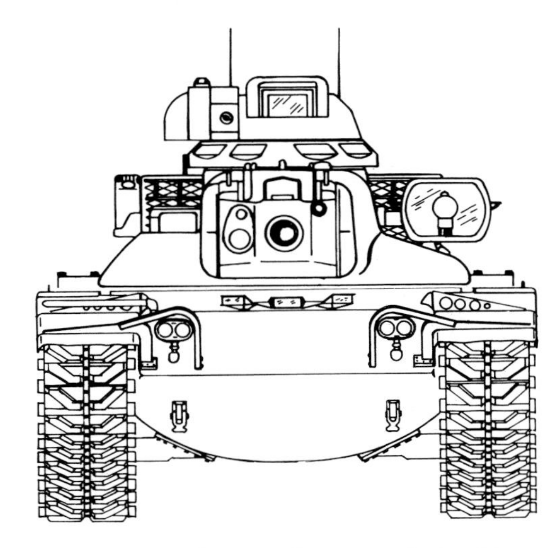

## SPECIFICATIONS

### 152MM Gun Tank M60A2

| | |
|---|---|
| **Crew** | 4 men |
| **Length (gun forward)** | 288.7 inches |
| **Width** | 143.0 inches |
| **Height (over cupola periscope)** | 130.3 inches |
| **Ground Clearance** | 15.3 inches |
| **Combat Weight** | 114,400 pounds |
| **Armament** | |
| **Main** | 152MM Gun-Launcher |
| **Secondary** | |
| **Cupola Mount on Turret** | 50 Caliber Machine Gun |
| **Co-Axial** | 7.62MM Machine Gune |
| **Engine** | AVDS-1790-2A |
| **Horsepower** | 750 HP |
| **Maximum Speed** | 30 MPH |
| **Maximum Range** | 280 Miles |

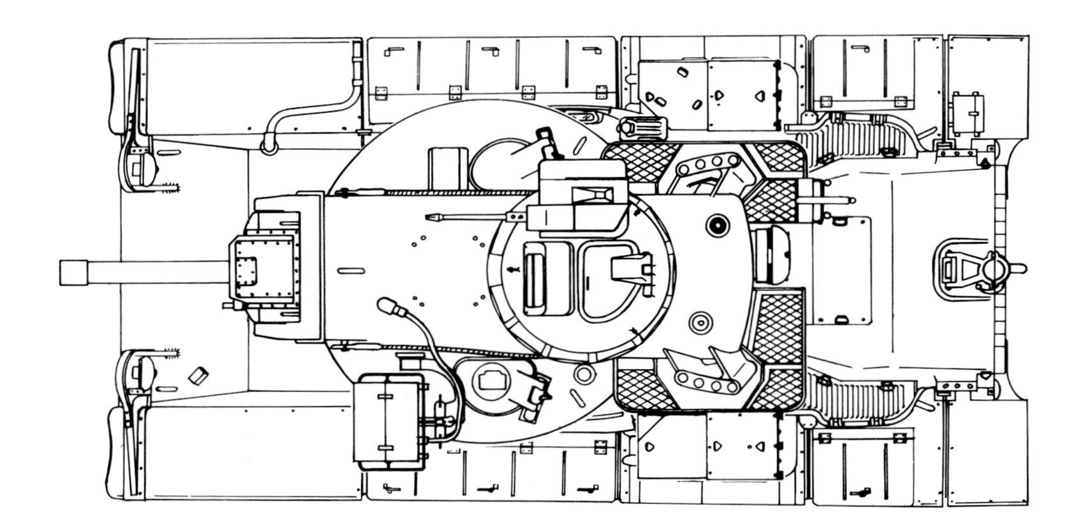

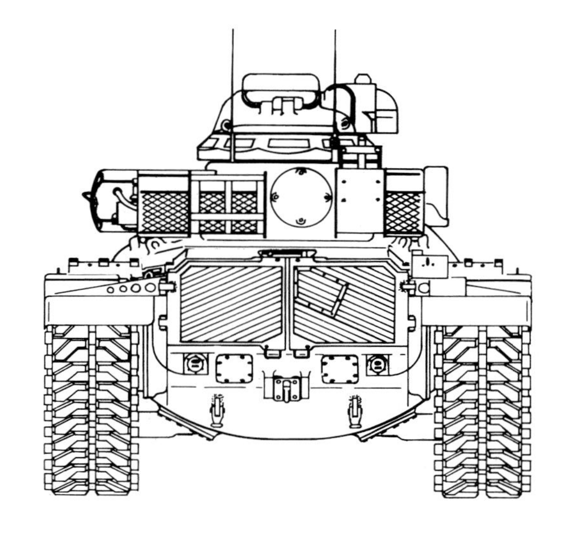

# M728 Combat Engineer Vehicle (CEV)

The M728 Combat Engineer Vehicle (CEV) was a specially designed variant of the M60 series produced for use by combat engineering units. The mission of combat engineering units called for the removal of obstacles which impeded the advance of armored forces, obstacles which included roadblocks, enemy bunkers, or damaged vehicles. The removal of such obstacles, often carried out under enemy fire, called for an armored vehicle equipped with specialized equipment designed specifically to remove these obstacles.

The M60 combat engineering vehicle, under the designation M728, was developed as a result of the Army's decision to cancel a similar vehicle based on the experimental T95 tank design then under development. This experimental CEV vehicle, designated the T118, had undergone tests beginning in 1959, but when the Army decided to adopt the M60 as its main battle tank, the Army decided to abandon the T118. However, the T118 prototypes were used to test the engineering equipment that would later be used on the M60 chassis. But, while the bugs in the new equipment were gradually worked out, it was found that the T118 turret could not be mated to an M60 hull without extensive modifications, consequently the army decided to mount the engineering equipment on an M60A1. The modified tank received the designation CEV T118E1.

The CEV T118E1 had three major modifications which distinguished it from a regular M60A1. The main armament installed was a 165MM gun based on a British L9A1 cannon, but without a bore evacuator or the semi-automatic breech mechanism. It had the breech bushing integrated with the gun tube and was fitted with a modified breech block latch, obdurator, and firing mechanism. The entire system was fitted into a modified M60A1 gun mantel. Due to the size of the ammunition and engineering equipment only thirty rounds of ammunition could be carried by the tank.

The second major feature of the T118E1 was the addition of an 'A' frame boom mounted on the turret, which pivoted from a point just behind the gun mantle and could be raised or lowered from inside the turret. With a single-part line the frame could lift nearly nine tons and while not strong enough to lift a heavy vehicle it was considered more than adequate to remove most obstacles.

The third major feature of the T118E1 was the addition of an M9 bulldozer blade mounted on the front of the vehicle. This dozer blade could be used to push aside stalled tanks, remove roadblocks, or dig emplacements for other tanks so they could take advantage of a defensive hull down position. The CEV T118E1 was also equipped with a 25,000 pound capacity winch in the turret bustle for towing or removal of mired down vehicles.

By 1966 the vehicle was ready for issue to armored units. Under the production designation M728 Combat Engineer Vehicle the new tank was issued to armored engineering units both in the United States and overseas. Later, in the 1970's the M728 began to filter into reserve and National Guard units as these formations were upgraded to meet an expanding role in US strategy. The new CEV saw limited combat during the later stages of the Vietnam War and performed well under the harsh climatic and terrain conditions found in Southeast Asia. There is an unsubstantiated report that an M728 engaged and destroyed a communist T54 with its 165MMhowitzer, but has yet to be confirmed. Other than this service, the CEV has been involved in no other combat operations, and other than US forces, has only been supplied to Saudi Arabia. At the present time there are no plans to replace the M728, even with the introduction of the M1 Abrams. The army feels that the M728 can carry out its intended role into the early 1990s and is in no hurry to procure a replacement.

**(Above) The first production line model of the M728 demonstrates its power by lifting the turret casting of an M60A1 at a press demonstration by Chrysler. Chrysler produced 243 of these vehicles at their Detroit facility beginning in the summer of 1966. (Chrysler via Binder)**

**(Below Left) The pivot mechanism and hydraulic arm on the right side of the turret. A standard M19 cupola was carried on the M728. (Mesko)**

**(Below) The A-frame is in the stowed position with the turret traversed to the rear. (Mesko)**

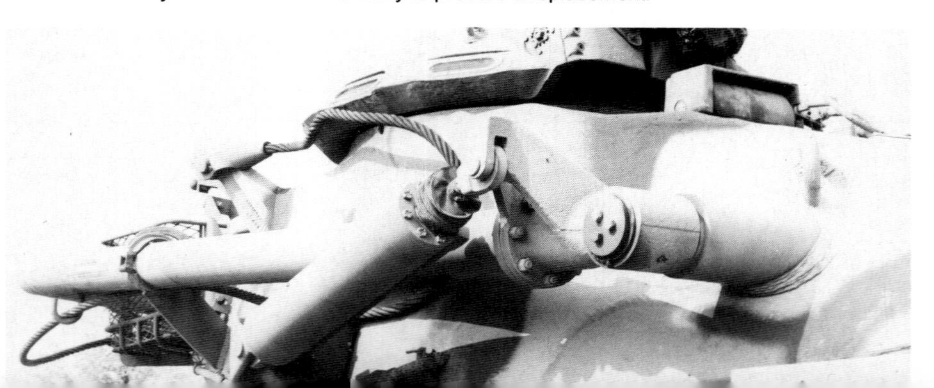

(Above) The rugged M9 bulldozer blade was made of heavy steel, braced with heavy steel fittings, and could be fitted to a standard M60 or M60A1. (Mesko)

(Above) The dozer blade used on the M60 series was adapted from the M8A1 dozer kit used on the M48 series and was a very reliable accessory. The headlights have been raised to overcome the height of the dozer blade. (Binder)

(Right) Aside from the dozer blade and A-frame the most prominent feature of the CEV is the short barreled 165MM demolition cannon which was a redesign of the British L9A1 cannon. (Binder)

(Below) The winch was fitted to the rear of the turret, forcing the standard storage rack to be cut down. The jerry can holder, which was normally mounted on the turret side, is now mounted on the end of the winch cover. (Mesko)

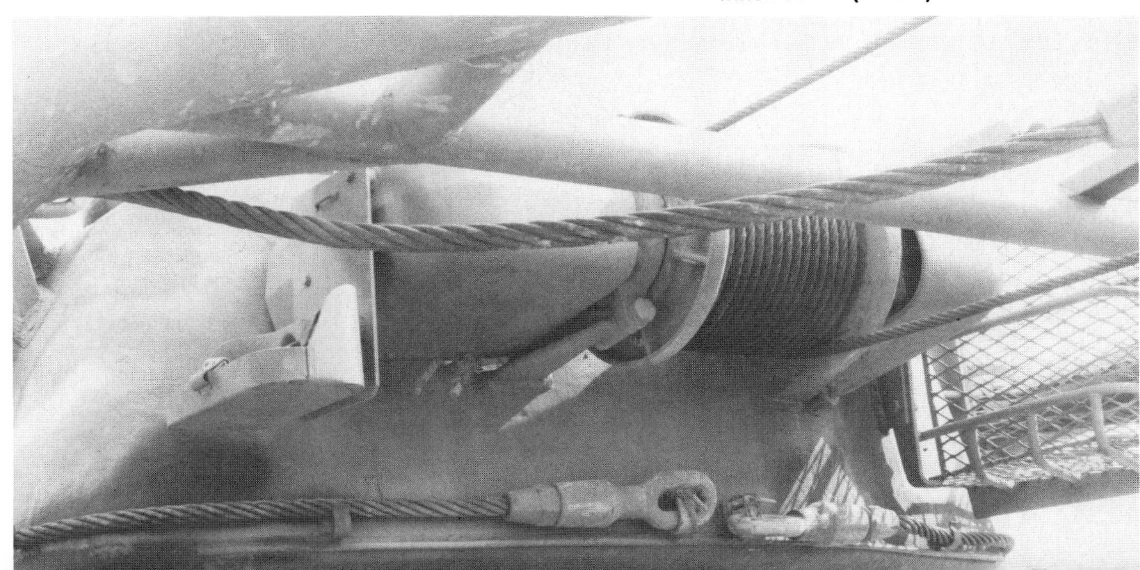

**(Above) A Seventh Army M728 carrying one of the first camouflage patterns adopted in the early 1970s for US armor in Europe. The supplementary arm attached to the pivot mechanism has been removed, possible for repair. (Seventh Army via Binder)**

**(Right) An M728 uses its winch and A-frame to remove an obstacle from a pathway during a winter exercise in Germany. The frame can lift up to 17,500 pounds. (Petersen via Binder)**

**(Below) A CEV is ferried across a river during maneuvers by the 3rd Armored Division during the fall of 1979. The M728 could be used in preparation of bridging sights and the establishment of tow lines at a river crossing. The soldiers are moving blocks under the tread to make sure it does not slip to the side. (3rd Armor via Binder)**

(Above) The dozer blade of the M728 can be employed for road clearing, digging emplacements, moving vehicles and a host of other duties. This CEV is using its blade to clear an obstruction on a dirt road in Germany. The raised headlights and guards can be seen very clearly. (Petersen via Binder)

(Above Left) Covered with equipment, a CEV of the 3rd Armored Division moves through a forest in Germany during maneuvers. The crew has used the A-frame to hold gear as well as attaching a pack to the searchlight bracket. (3rd Armored Division via Binder)

(Left) A CEV often travels with its turret traversed to the rear. This M728 climbs an embankment during a training session at Fort Riley, Kansas. The headlight guards have been removed from this vehicle. (US Army/1st Infantry Division via Binder)

# M60AVLB

With the introduction of the M60 series in 1960, the army moved to standardize as much equipment as possible within its regular armored formations. As a result of this decision, the M60 chassis replaced the M48 as the chassis of the armored vehicle launched bridge (AVLB). The M60AVLB began reaching regular armored formations in 1963 and plans were drawn up to eventually replace all of the older M48AVLBs with the M60AVLB. M48AVLBs were gradually withdrawn from active Army units and reassigned to reserve or national guard formations. However, the plan to replace all of the M48 bridge layers was later modified and M60 components were used to rebuild M48 bridge layers to M60 standards instead of procuring new vehicles, saving a substantial sum of money.

Production line change-over from the M48 to the M60 caused no major problems since the basic bridgelaying equipment of the M48AVLB was used on the M60AVLB with only minor modifications, mainly in the mounting and support points on the M60 hull.

The bridge itself came in either 43 foot or 63 bridge foot sections, with the 63 foot section being the most common. The bridge unit is made up of symmetrical halves which are hinged together, and folded atop one another when at rest on the tank. The spans, made of aluminum and steel, weigh approximately fourteen tons and increased the height of the vehicle by two feet over the regular M60. The bridge rests on a steel girder located over the rear hull with the launching mechanism mounted on the front glacis plate. When activated, a hydraulic cylinder pushed the bridge upward while the launching mechanism swung downward tipping the bridge unit forward. As the mechanism moved the bridge unit forward it plants a plate in front of the tank which gradually absorbs the weight of the bridging span. Another cylinder eases the bridge down and across the obstacle until it is in place. Once the span is in place it can be disengaged by an ejection cylinder, leaving the vehicle free to move away while the bridge is used by other tanks. The the AVLB can then reconnect to the span and raise it back into the carrying position. Since both ends of the unit have connecting points, the vehicle can recover the bridge from either side of an obstacle. Under ideal conditions it only takes two minutes to lower the span and about ten minutes to recover it, but this does not take into consideration lining up the unit at an obstacle to be bridged.

The M60AVLB has been supplied to US allies throughout the world. It was used in combat by American forces in Vietnam alongside the M48AVLB where it performed reasonably well under the difficult weather and terrain conditions. It is not believed that any were supplied to the South Vietnamese army when US forces pulled out in 1973. The Israelis used the M60AVLB during the 1973 Yom Kipper War when they crossed the Suez Canal. While the span was not wide enough to bridge the waterway, the unit proved extremely useful as a replacement for pontoon sections damaged by Egyptian gun fire and for bridging the gap between the pontoons and the shoreline. The Israelis also used their AVLBs during the invasion of Lebanon to span bridge sections destroyed by Palestinian guerrillas attempting to slow down the Israeli armored thrust.

At the present time tests are underway to develop a longer and heavier bridgelaying unit for use on the M60 chassis. This new unit will hopefully be able to handle the heavier equipment which may come into service during the next decade. There are also plans under study to develop a new AVLB version based on the M1 Abrams chassis. In the meantime the current M60AVLB will remain the standard bridgelaying equipment in regular army units, and whatever comes of these plans, it will be some time before the M60AVLB is retired from active service.

**(Below) Except for relatively minor fittings, there was little difference between the AVLB equipment of the M48 and M60. An employee of the Rock Island Arsenal mates the hull and launching mechanism of an M60 AVLB. The flat portion of the plate served as a base when the bridge was launched. (Rock Island Arsenal via Binder)**

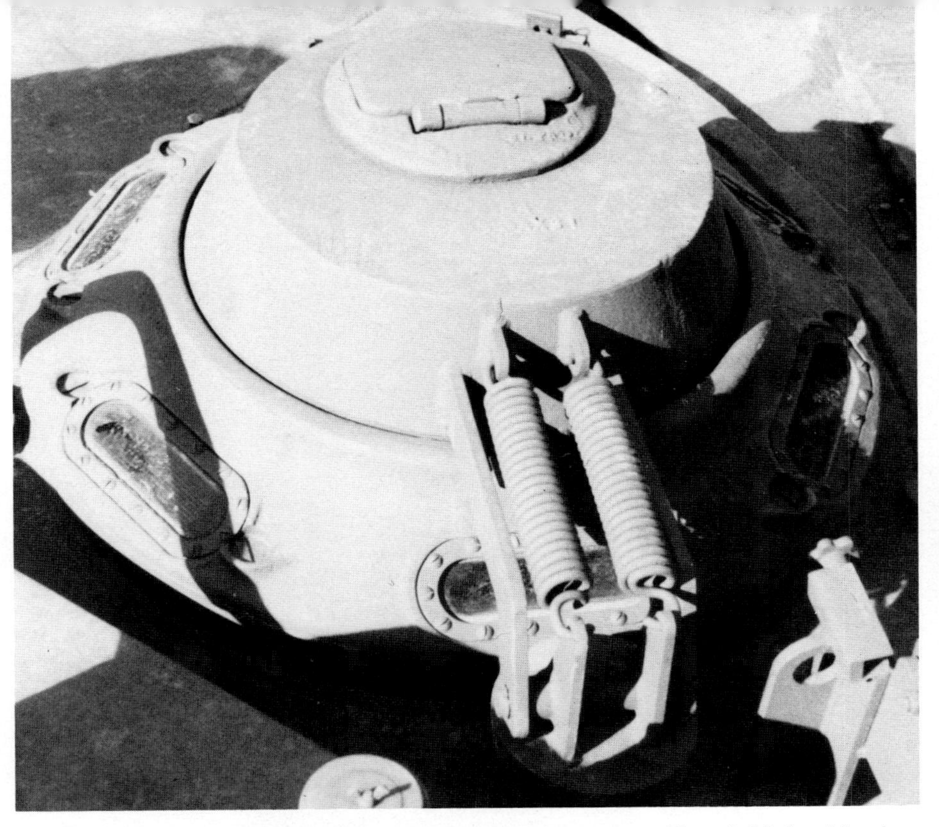

(Above) As in the M48 AVLB the driver and commander were positioned side by side where the turret would have been on a regular tank. (Binder)

(Below) The I beam that the bridge rests on can be seen at the rear of the engine deck. These hatches were peculiar to the AVLB variant of the M60. (Binder)

(Above) The AVLB without the bridge atop it. The I beam on the engine compartment holds the bridge when the unit is being carried. The older style side loading air filters can be seen next to the engine. (Binder)

## Bridge Span 60 Feet

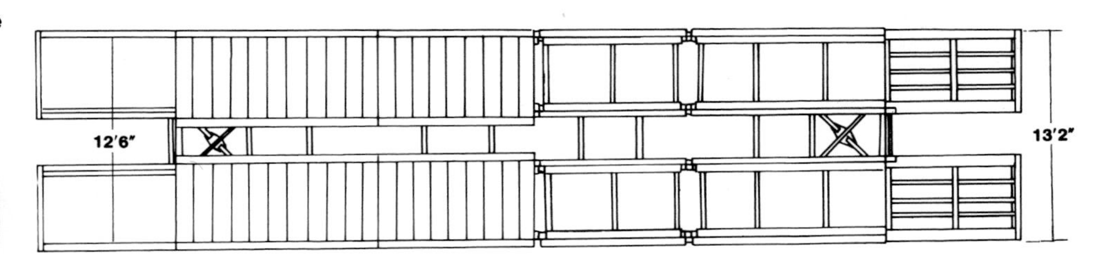


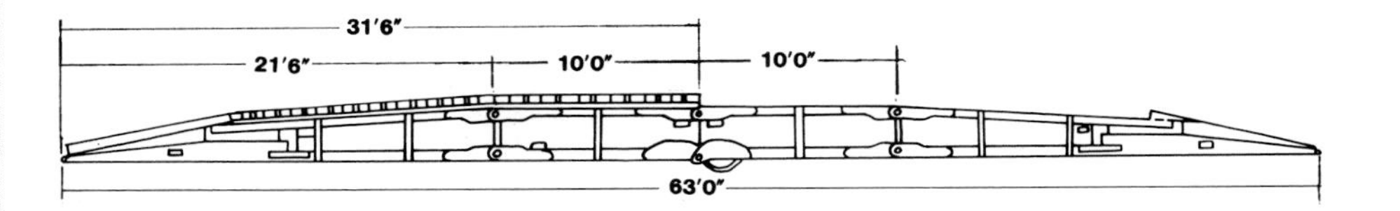

(Above) A bridge partially raised prior to emplacement during a demonstration. The base plate has yet to make full contact with the ground. (US Army via Binder)

(Above) The launching mechanism and attachment points. Once the bridge is in place the tank could be disengaged in two minutes, allowing vehicles to proceed over the bridge span. Under normal condition the AVLB needed about ten minutes to reattach itself. (Binder)

(Right) The launching mechanism and boom assembly have been lowered and the base plate has plowed up the ground around it to form a firm base on which to pivot the spans. (Binder)

(Left) The bridge has been lowered before being placed on the ground. The plate is now flat on the ground and provides a stable base from which the unit may be lowered to the ground. (Binder)

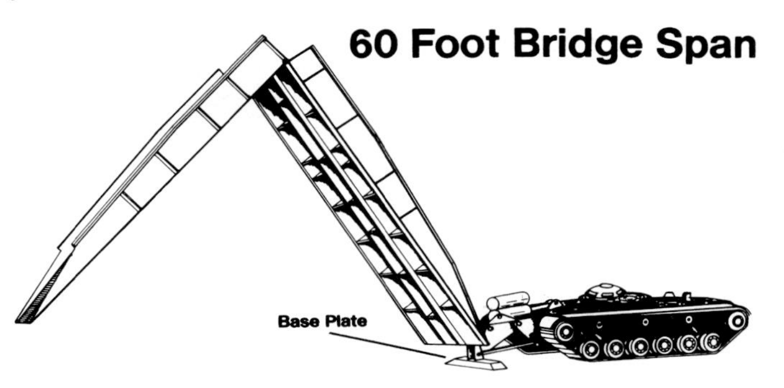


(Below) The bridge is being extended and lowered to span a small stream during 'Exercise Reforger III' in Germany during 1971. (US Army via Binder)

(Above) As the bridge nears the horizontal extended position the two spans gradually come together. Once fully extended pins are placed in the upper hinges of the spans to secure them to each other. (US Army via Binder)

(Below) Looking down into the driver/operators position. The levers extending down from the top are used to control the action of the bridge during emplacement. (Binder)

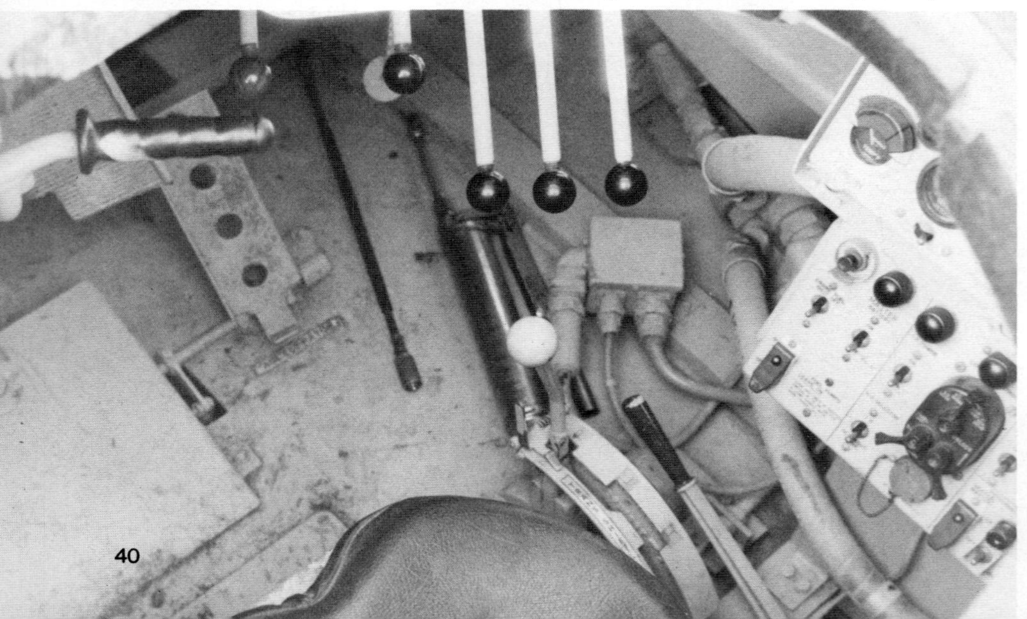

(Below) With the bridge unit locked securely on the chassis the M60 AVLB is only two feet higher than the normal gun tank version. This particular AVLB is camouflaged in the four tone MERDC scheme developed by the Army during the 1970s. (4th Infantry Division via Binder)

# Combat Operations

## Vietnam

That the M60 series has seen far less combat service than its sire, the M48, is only partially due to its newness, but also because it has not been exported as widely as the M48. The first M60s to see combat were the M728CEV and the M60AVLB which saw action with US forces in Vietnam on a limited scale. Used in support of US and Vietnamese armored and infantry units both vehicles functioned extremely well. In particular the M60AVLB proved extremely valuable during operations in I Corps where they were used to replace the destroyed bridges over the streams which crisscrossed the rolling mountainous terrain, and also played a vital role in operations around the Khe Sanh area where the communists had destroyed most of the bridges in order to impede the movement of allied armor. There is an unsubstantiated report that an M728 destroyed a North Vietnamese T55 in an engagement but little is known about this incident. While there are also accounts of M60s being sent to Vietnam for test purposes, but this seems unlikely, since no photos or records have yet been unearthed to support the reports. The Army felt that the M60 offered no significant advantage over the M48 except for its 105MM gun, and felt that the 90MM gun of the Patton would be more than sufficient for requirements in Vietnam especially since there was little possibility of tank versus tank combat.

## Yom Kipper War

The first test of the M60 series in tank versus tank combat came during the 1973 Yom Kippur War between Israel, Egypt, and Syria. In the mid 1960s Israel received a shipment of M48s from Germany with American approval. And while they liked the M48, the Israeli Army felt it was underarmed and began fitting a 105MM gun. Some of these up-gunned M48s were used in the 1967 Six Day War where they performed exceptionally well. As a result Israel ordered additional M48s and both the M60 and M60A1 to off-set the increased Soviet aid following the Arab's defeat in 1967.

On 6 October 1973 the Arabs, frustrated with both the political and military status quo, launched a surprise attack on Yom Kippur, the high Jewish holiday. The bulk of the M48s and M60s were stationed along the Suez Canal opposite the Egyptian forces. When the Egyptians launched their attack over the waterway against the fortified observation points of the Bar Lev line, Israeli armor was rushed forward to support these positions and bring the enemy assault boats under fire. Unfortunately, the Israeli armored advance was made without adequate artillery or infantry support and ran into hordes of infantry equipped with anti-tank missiles, recoilless rifles, and rocket propelled grenades (RPGs). In addition Egyptian artillery and tank fire from the opposite bank hit many of the advancing Israeli tanks. While some Israeli tanks managed to fight their way to the fortified points many were knocked out or forced to retreat by the heavy volume of fire.

During the first few days the Israelis were unable to dislodge the Egyptians who quickly built up their forces and expanded their bridgehead. The Israelis regrouped and launched a number of counterattacks, but again, these were not well supported by infantry or artillery and were repulsed with heavy losses. Stunned by these setbacks, and worried over events in the Golan Heights, the Israeli army dug in to contain the Egyptians until the situation in the north stabilized. Once the Syrians were contained, forces could be shifted from the north to push the Egyptians back across the canal.

During this digging in period the Egyptians launched a number of major attacks against the Israeli positions which were thrown back with severe losses to the Egyptians. As these battles were going on the Golan situation gradually stabilized and new equipment began arriving from the United States, including M60s to replace the armor lost during the early fighting. By mid-October the Israelis, reinforced and reequipped, were ready to launch their counterattack. Striking at a gap between two Egyptian armies, the Israelis broke through to the canal near the Great Bitter Lake on 7 October. Crossing the canal on rafts, Israeli troops secured a bridgehead and tanks were quickly ferried across to reinforce them. At first, the Egyptians, unaware of the threat, did little to hinder the expanding Israeli bridgehead. However, fierce fighting took place around the eastern perimeter, especially around a position called the Chinese Farm. For three days and nights both sides slugged it out. Tanks were often only a few meters apart. Finally, the Israeli forces won out and pushed the Egyptians back.

Meanwhile, Israeli engineers had constructed a pontoon bridge over the canal and an entire armored division was moved over the canal to disrupt the enemy lines of communication. M60AVLBs were used to bridge the gap between the bridge and the shoreline, and to replace decking lost when pon-

**(Above) The only variants of the M60 to see service in Vietnam were the bridgelayer (AVLB) and the combat engineer vehicle (CEV). This M60 AVLB is taking part in OPERATION UTAH MESA near Khe Sanh where some of the fiercest fighting of the war took place. Many bridges and roads were destroyed by the enemy to hinder allied armored units, and the bridgelayers performed valuable service in support of armor during relief and clearing operations. (US Army)**

**(Below) The M728CEV saw limited service with engineer battalions in Vietnam. The crew of an M728 work on the vehicle's track at a base camp near Quang Tri. The CEV is from A Company, 7th Engineer Battalion.(US Army via Binder)**

**(Above) An Israeli M60 climbs an embankment in occupied Jordanian terrain near Jerusalem. A variety of markings have been painted on the tank and on the canvas panels attached to the turret. (IDF via Binder)**

**(Below) Following the performance of the M48 in the 1967 War Israel decided to purchase more of them along with the newer M60 and M60A1. These M60s take part in a parade just prior to the 1973 Yom Kippur War. The Israelis have removed the M19 cupola and replaced it with a low profile cupola. (IDF via Binder)**

toons were destroyed by enemy artillery fire. At the same time M48s and M60s were busy destroying Egyptian missile batteries, disrupting supply lines, and repulsing attacks on the bridgehead. By 22 October the Egyptians had had enough and requested a ceasefire. The fighting finally stopped for good on 25 October when UN forces separated the two sides.

Though losses were heavy on all sides the Israelis came out the firm victors on both fronts. Despite their early tactical mistakes along the Suez Canal, the Israelis were able to quickly modify their battleplans to compensate for the large numbers of Egyptian anti-tank weapons. In tank versus-tank action the superior training of the Israeli tank crews often proved to be a decisive edge in confrontations with the Russian supplied T-54s, T-55s and T-62s. In particular the Israelis were very pleased with their M60s and M60A1s although they did discover a number of weak points in the design. One serious problem arose when the M60A1s were hit in the chin area under the turret front or the turret ring. These areas proved to be underarmored and were penetrated fairly easily, resulting in the total loss of the tank. The hydraulic fluid used in the turret traverse mechanism also caused the loss of a number of tanks because of its low flash point which ignited with relatively light battle damage. The Israelis ordered more of the tanks and supplied information to the United States which helped correct the flaws.

## Into Lebanon

Following the Yom Kipper War in 1973, Israel experienced an increase in terrorist raids from Lebanon. The Lebanese Army, unable to counter these raids, disintegrated as a viable fighting force. Syrian forces moved into eastern Lebanon which also posed a threat to the Israelis. As a result the Israeli army launched a number of raids into Lebanon to wipe out guerrilla bases, but despite these forays the terrorists continued their attacks with Syrian aid.

Frustrated with this situation, the Israeli army launched a full scale invasion of Lebanon in June of 1982 with over 1400 M60A1s, Centurions, and the new Merkava tank. Initially, the Israeli's three pronged advance encountered light resistance suffering minimal casualties, mostly to RPG's or anti-tank missiles. However, as the Israeli forces pushed north, the Syrians moved to counter the Israeli thrust. The two forces met on 10 June south of Beirut and fought a pitched two day battle. By the end of the second day Syrian forces were in full retreat. Over 400 T-55s, T-62s and T-72s lay smoldering on the battlefield while only some fifty Israeli tanks were destroyed. Using a new projectile designed by the Israeli Military Industries, the Israeli tankers were able to engage the Syrian tanks at ranges over 2000 meters and still penetrate the frontal armor with devastating results.

Following this victory the Israeli advance continued until Beirut was invested. International pressure, however, forced a halt to the offensive just when the decisive phase was about to begin, allowing the guerrillas to escape. Throughout the offensive the armored edge of the Israeli army had constantly kept unrelenting pressure on the guerrilla forces. An interesting modification to counter the various hand held weapons used by the guerrillas was experimented with. Explosive armor packs were fitted to Israeli tanks to defeat the high explosive anti-tank (HEAT) rounds used by the Arabs. Called 'Blazers', these packs were detonated when struck by the HEAT projectile. Upon detonation, the jet from the HEAT round was dissipated, thus destroying its effectiveness against the tank's armor. This effectively reduced the guerrillas ability to destroy Israeli armor with hand held weapons. Measured against the slight increase in weight, the Blazer proved worthwhile throughout the campaign.

Again the M60 proved to be extremely effective against the latest Russian armor. With the new projectiles the Israelis were able to destroy the new Soviet T-72 at extreme range despite its supposedly advanced armor protection. This new Israeli projectile has been supplied to certain NATO countries by Israel, and is similar to the American depleted uranium round in performance. The M60 series tank can still engage the newest Russian armor on equal terms.

## Other Conflicts

Other than the mid-East the M60 series has seen little combat. A few USMC M60A1s were used in the liberation of Grenada by American forces during the fall of 1983. The only other area where the M60 has been used is in the war between Iran and Iraq. Iran had acquired over 400 M60A1s during the reign of the Shah. During the turmoil following the Shah's downfall Iraq launched an offensive to retake an area ceded earlier to Iran. While little factual information is available on the fighting, it is known that armor has been used extensively. Losses on both sides have been extensive on both sides, with neither combatant exhibiting any flair for using their armored forces. Little information of value has been gained about the respective performance of the vehicles on either side.

(Left) A brand new M60A1 stands ready for inspection in the Sinai a few months before the Egyptian attack. The Israelis repainted their vehicles in a Sand-Gray color which helped them blend into the surrounding terrain. (IDF)

(Below) With the surprise Egyptian attack in October of 1973 Israeli armor once again faced off against Russian supplied equipment. This M60A1 moves up toward the front to counter the initial Egyptian assault across the Suez Canal. (IDF)

(Above) Unfortunately the initial Israeli counterattacks lacked artillery or infantry support and suffered severe losses to Egyptian armored and infantry forces. This partially burned out M60A1 lost a track to a mine and then caught fire. (Egyptian Embassy via Binder)

(Below) The Israelis were able to move up supporting forces and finally stopped the Egyptian advance. This M60 maneuvers into position to help repulse an Egyptian armored thrust during the second week of the war. (IDF via Binder)

(Above) Fierce fighting continued to take place, sometimes even at ranges of only a few yards, and while Egyptian losses were high, the Israelis also suffered heavy losses. A weak point which was discovered in the M60 design was the use of a turret hydraulic fluid which caught fire easily when a line was ruptured. This M60A1 was completely destroyed by such a fire. After the war a fluid with a high flashpoint was substituted. (Egyptian Embassy via Binder)

(Above) The incursion by Israeli forces into Lebanon brought them into direct conflict with Syrian armor. This M60 fires on a Syrian position south of Beirut. The objects attached to the hull are explosive armor packs designed to defeat HEAT rounds. Most if not all M60s used in the Lebanon invasion were fitted with these. (IDF)

(Left) To make up for Israeli losses the United States rushed large numbers of M60s from stocks in Europe to replenish Israeli armored units. This M60A1 is being unloaded from a C141 at Tel Aviv. This emergency resupply effort seriously depleted US tank strength and resulted in the updating of the older M48 fleet until production could offset the losses. (US Air Force via Binder)

(Below) This M60 has been fitted with a plow to dig up mines in Lebanon. The many attachment points welded on the turret and hull are for mounting the explosive packs. (W E Storey)

# M60A3

The M60A3 is the final version of the M60 series slated for production. This variant, first introduced in 1978, is based on the upgraded M60A1 which had been progressively modified during the early 1970s. These progressive changes made to the M60A1 resulted in a far superior vehicle when compared to early M60A1 variants and led the Army to give it its own designation. The most significant additions were a ruby laser rangefinder and a solid state ballistic computer which tremendously increased the probability of a first round hit. Additional changes included the addition of cheek fillets to the forward part of the turret, a new M240 coaxial machine gun, extra armor on the turret ring, British style smoke dischargers mounted on the turret, and a thermal sleeve on the gun barrel to prevent heat warpage.

Shortly after the introduction of the M60A3, another modification led to a sub-variant under the designation M60A3TTS in which the standard M35E1 passive night sight was replaced by an AN/VSG-2(TTS) Tank Thermal Sight. The TTS picks up infrared emissions from objects in the line of sight and amplifies it into a coherent image for the tank commander or gunner. Since all objects give off a certain amount of heat or infrared energy, this sight can distinguish between various objects by the level of energy it picks up. As such, it provides the tank crew with the ability to pick out an enemy vehicle in the dark, through mist and fog, and battlefield conditions where there is a great deal of smoke and dust. The M60A3 was the first production line tank to be outfitted with such a device which gives it a decided advantage over other tanks. In addition the M60A3 also has a meteorological sensor which picks up wind velocity and feeds it directly into the ballistic computer, allowing corrections to be made and further increased the chance of a first round hit. Finally, a smoke generator was added to the engine which could be used to create a smoke screen during battlefield movements. This smoke screen generator is similar to one used in Soviet tanks and adds another dimension of camouflage which can be used by the crew to increase their chances of survival in combat.

Compared to current Russian tanks, the M60A3 comes out favorably. More heavily armored than the T55 or T62, it may not be as well armored as the T64 or T72. However, with the new M735A1 hyper-velocity discarding-sabot armored-piercing round which uses a depleted uranium core, the 105MM gun has the ability to penetrate the T64 or T72 armor at long range. This was well illustrated in Lebanon when Israeli M60s engaged Syrian T72s and dispatched them with little problem. In terms of mobility, the later Soviet tanks have a somewhat higher speed but are harder to steer and do not have as reliable a transmission as the M60. The fire control system in the M60A3 gives the crew a better chance of a first round hit than Soviet fire control systems, which, when coupled with a higher rate of fire and a larger ammunition load, the M60 has a definite advantage under most battlefield conditions. Finally, while all Soviet tanks have a lower profile than the M60, the higher US tank is roomier and puts less of a strain on the crew. With a lower fatigue factor American tank crews can theoretically operate under combat conditions for a longer period of time at peak efficiency. The higher M60 profile also allows the main gun to depress lower which allows the M60 to engage enemy tanks from a lower hull down position and be less exposed than their Soviet counterparts. Even though, the basic M60 design is over twenty years old and is based on an even older design, the M60A3 TTS is quite capable of holding its own against the current field of Soviet armor. Future developments in tank technology will eventually make the M60 obsolete but for the foreseeable future it will remain a viable weapons system and the will remain the American main battle tank until sufficient numbers of the new M1 Abrams are available.

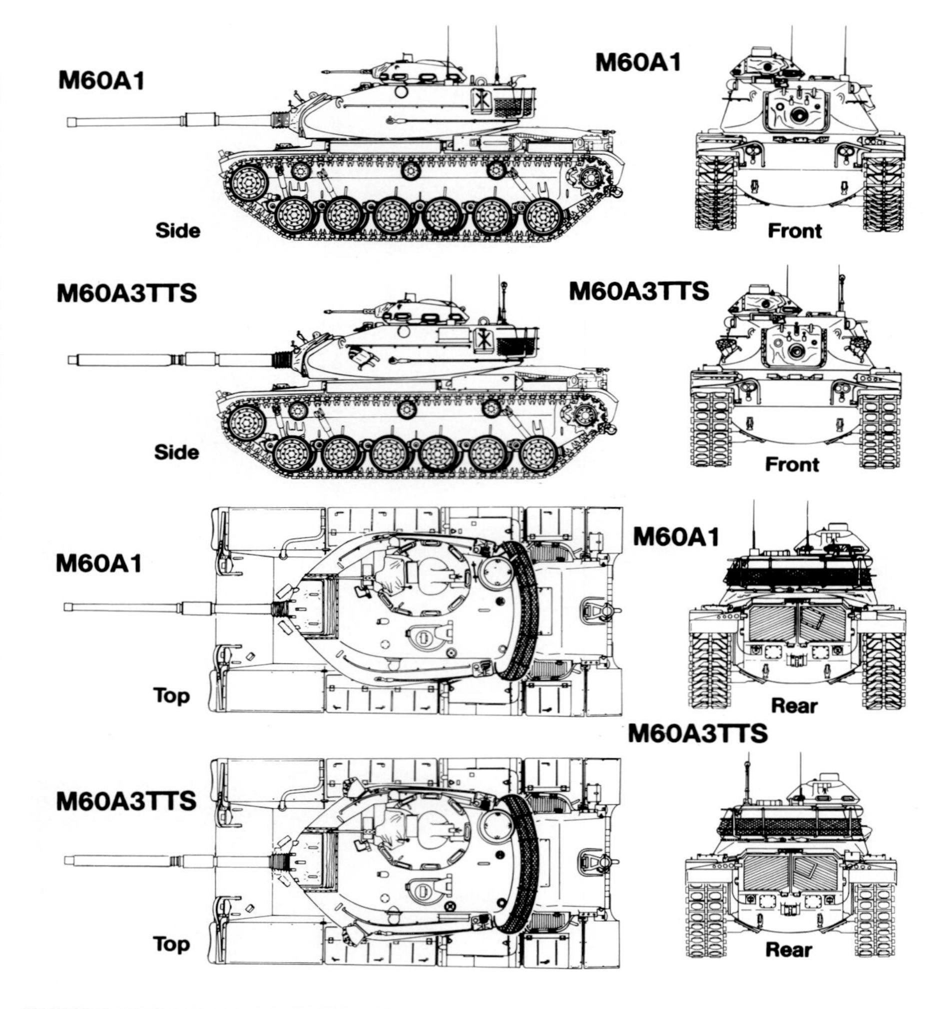

**(Right) One of the M60A1E3 pilot vehicles used to test the various components which eventually resulted in the M60A3. External changes included the chin armor, top loading air filters, cross wind sensor, new tracks, and smoke launchers (not carried on this vehicle). (US Army via Binder)**

(Above) The smoke grenade launchers were adapted from a British pattern used on the Chieftain. In US service these grenade launchers are designated the M239. (Chrysler via Binder)

(Above Right) One of the more subtle ways to tell the M60A3 model from upgraded M60A1s is by the ranger finder installed on the right side of the turret which has an armored flap cover. One of the first M60A3s to arrive in Germany, this vehicle has the range finder flap cover in the open position. (Seventh Army via Binder)

## M239 Smoke Discharger

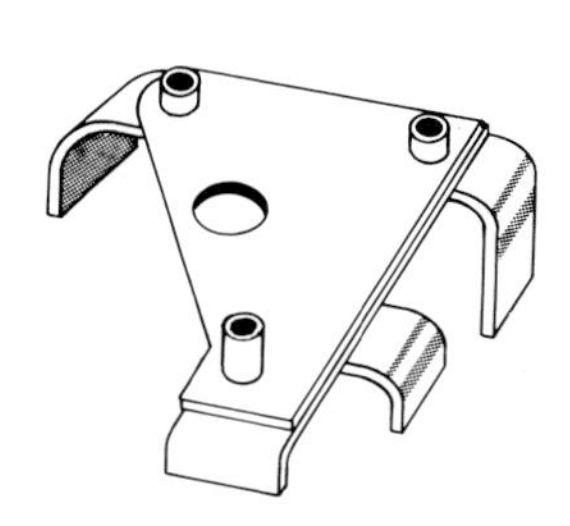

Attachment Rack

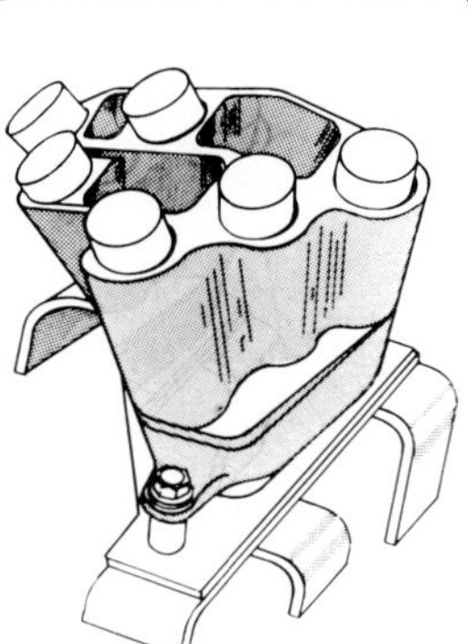

Smoke Discharger

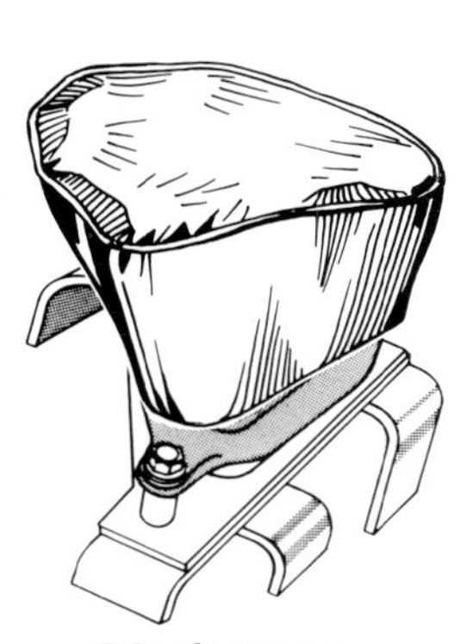

Discharger with Cover

(Right) The four tone MERDC scheme has become the standard camouflage pattern for the M60 series in Europe although at the present time a three tone scheme is under consideration as a possible replacement. Note the thermal guard on the gun barrel and the pattern of the T142 track shoes. (US Army via Binder)

(Above) This M60A3 has had the forward thermal guard removed to make it easier to mount the gunfire simulator on the barrel. The thermal guard helps prevent barrel warpage due to the heat caused by prolonged firing. (3rd Armored via Binder)

(Above) *BUSTIN LOOSE* from the 32nd Armor, 3rd Armored Division during training exercises at Hohenfels, Germany. The large numbers are for identification of tanks which have been destroyed by the 'enemy' during such exercises. Only one round is left in the gunfire simulator mounted on the gun barrel. (3rd Armor via Binder)

## 105MM Gun Barrel

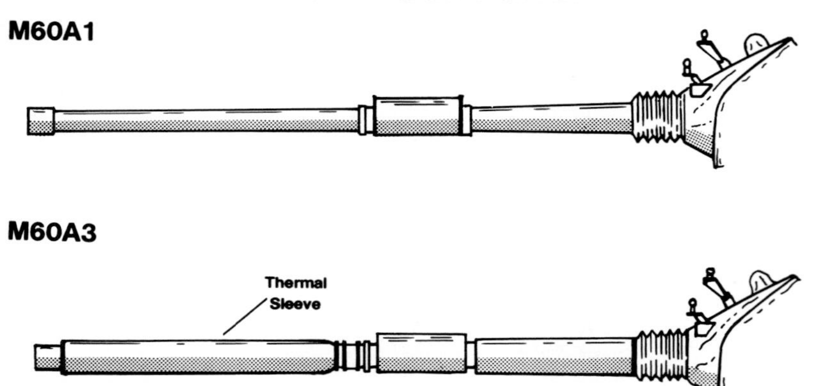


(Right) Serious threats to armored forces are chemical and biological warfare weapons, and armored forces must train constantly for the possibility of their use. This crewman washes down his tank as part of an exercise to counter the effects of such weapons. (3rd Armor via Binder)

(Above) Even with its high profile the M60A3 still possesses excellent armored protection and a hard hitting gun, which, when coupled with its new rangefinding equipment, makes it a serious threat to even the newest generation of Russian Armor. (3rd Armor via Binder)

(Right) It is becoming more and more of a common practice for tank crews to name their mounts with names ranging from 'gung ho' types of names to comic strip character names. The crew of this tank have named their M60 after the rock group Black Sabbath whose name appears on the bore evacuator. (1st Armored via Binder)

(Above) With the possible changeover to a new three tone camouflage pattern in the near future, the late production run M60A3s are not being camouflaged by the troops in Europe. Some tank crews are extremely skeptical about the value of camouflage schemes since tanks are so quickly covered with dirt and mud that the patterns are effectively covered up. (3rd Armored via Binder)

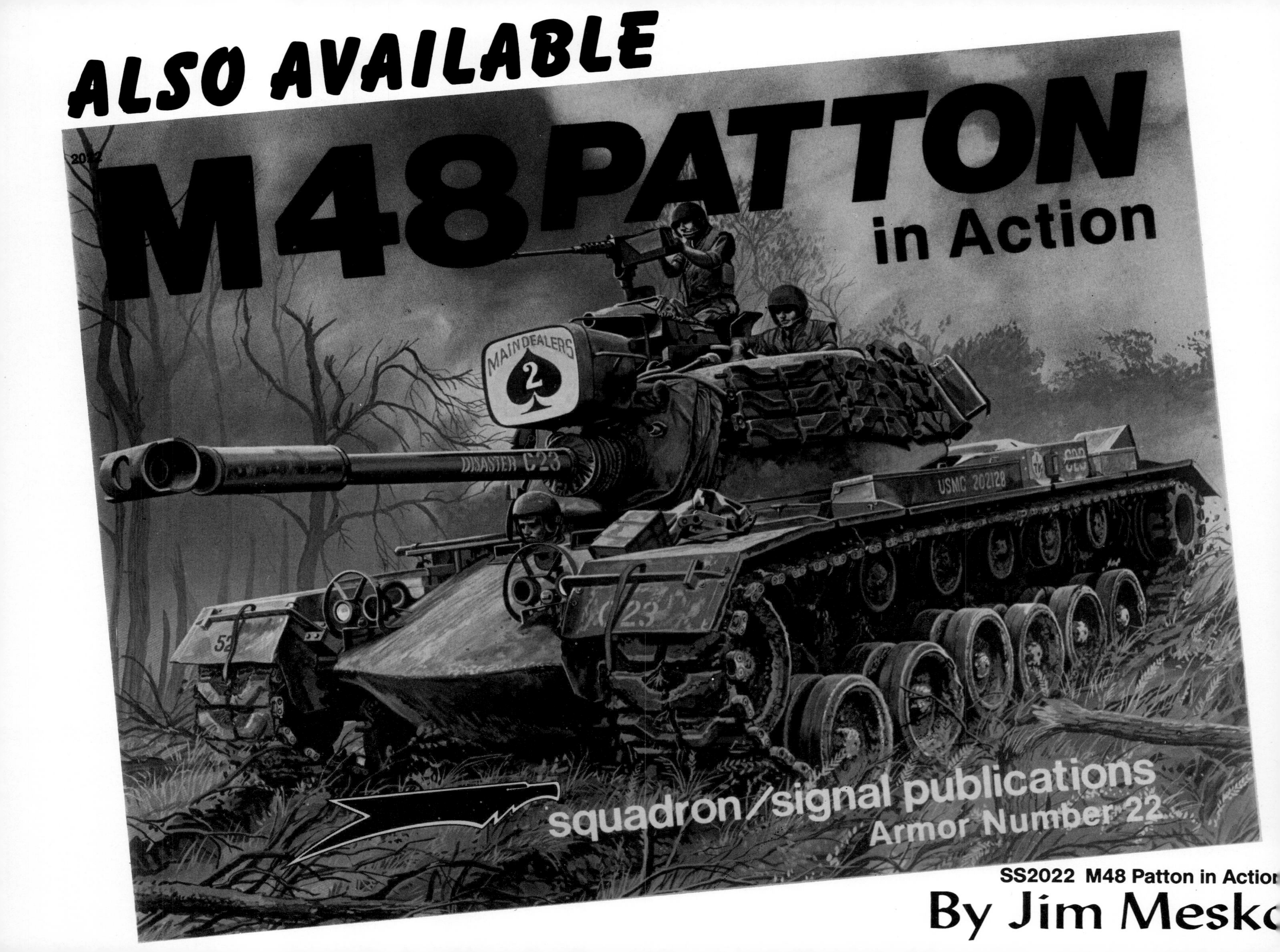
ALSO AVAILABLE
M48 PATTON
in Action
MAIN DEALERS
2
DISASTER C23
USMC 202128
C23
squadron/signal publications
Armor Number 22
SS2022 M48 Patton in Actio
By Jim Mesk